Becoming a STEM Teacher:
My Learning Journey

Becoming A STEM TEACHER

My Learning Journey

Dr Tan Aik Ling
&
Dr Melissa Neo

illustrated by Liu Yuxi

Published by

World Scientific Publishing Co. Pte. Ltd.

5 Toh Tuck Link, Singapore 596224

USA office: 27 Warren Street, Suite 401-402, Hackensack, NJ 07601

UK office: 57 Shelton Street, Covent Garden, London WC2H 9HE

Library of Congress Cataloging-in-Publication Data
Names: Tan, Aik-Ling, author. | Neo, Melissa, author.
Title: Becoming a STEM teacher : my learning journey / Dr. Tan Aik Ling &
     Dr. Melissa Neo ; illustrated by Liu Yuxi.
Description: New Jersey : World Scientific, [2024]
Identifiers: LCCN 2023039958 | ISBN 9789811284458 (hardcover) |
     ISBN 9789811284960 (paperback) | ISBN 9789811284465 (ebook for institutions) |
     ISBN 9789811284472 (ebook for individuals)
Subjects: LCSH: Science--Study and teaching (Primary) |
     Technology--Study and teaching (Primary) | Engineering--Study and teaching (Primary) |
     Science--Study and teaching (Secondary) | Technology--Study and teaching (Secondary) |
     Engineering--Study and teaching (Secondary)
Classification: LCC Q181 .T26 2024 | DDC 372.35023--dc23/eng/20231127
LC record available at https://lccn.loc.gov/2023039958

British Library Cataloguing-in-Publication Data
A catalogue record for this book is available from the British Library.

For any available supplementary material, please visit
https://www.worldscientific.com/worldscibooks/10.1142/13629#t=suppl

Desk Editor: Lum Pui Yee
Illustrator: Liu Yuxi

Table of Contents

AUTHORS' INTENTION

The road to becoming an integrated STEM teacher is a journey. As teachers, we oftentimes adopt different identities. Some are more general (like a form teacher), while others are more specific (like a science or mathematics teacher). These identities are interwoven into our teaching experiences and influence our self-confidence, self-knowledge and sense of being. In integrated STEM learning, we can similarly examine how our ideas of monodisciplines, like science, mathematics or technology can be featured when integrated with the learning and practices of other disciplines.

This STEM education journal aims to be a companion, or a guiding 'voice' if you will, for anyone who wishes to embark on this journey from *being* to *becoming* a competent STEM teacher.

Becoming an integrated STEM teacher requires time.

As you begin this book, you first engage in **being** an amateur integrated STEM teacher. Try out new ideas in the classroom, reflect on what you have done, retain the aspects and experiences you hold dear, and change or improve ideas that do not work.

Over time, you will transition from **being** an amateur integrated STEM teacher to **becoming** a confident integrated STEM teacher.

Each time you reflect, develop, and refine your practice, you transit to a higher level in your journey to becoming a better integrated STEM teacher. Transformation of practice can be arduous, scary, and sometimes lonely, often requiring multiple cycles of learning, practice, questioning, reflection, and refinement. As you progress through this book, it is our hope that you will discover that *being* and *becoming* a STEM teacher are transient processes, influenced by a constant refreshing of experiences and building of knowledge, much like the learning journey through your educational years.

There are two ways to use this book — **(1)** as a personal learning journey, and **(2)** as a companion for professional learning communities.

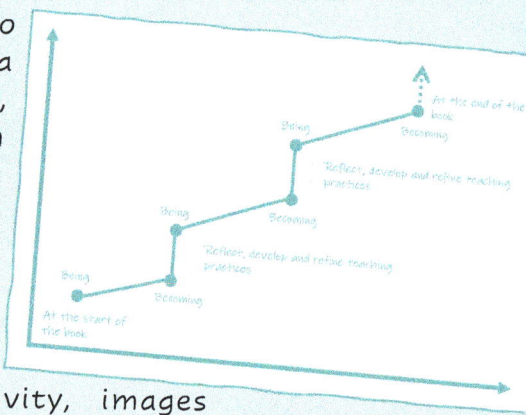

Six integrated STEM activities are presented as separate chapters, each with a probe activity, images of integrated STEM inquiry, an instructional model, and suggested instructional strategies.

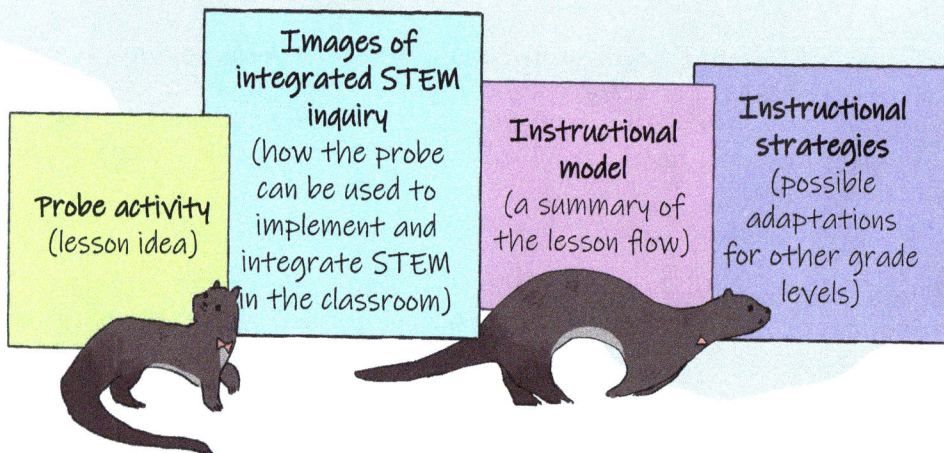

[Figure adapted from] Philosophy of being and becoming: A transformative learning approach using threshold concepts (6), by P. Natanasabapathy & S. Maathuis-Smith, 2019, Taylor & Francis.

Integrated STEM education requires disciplinary knowledge, pedagogy, assessment, and collaboration with others. As such, we have included questions for reflection, space for journalling and QR codes to access videos and other additional teaching resources, to cater to your classroom needs. For example, the Student STEM Workbook aims to increase the ease of implementing the integrated STEM activities featured in this book and help attain student feedback.

Each week, you or your team can read through the ideas presented in an activity and meet to discuss your insights and plans for implementation. After you have implemented the lesson, take some time to reflect, record your thoughts and have a follow-up discussion with your team.

We hope that this journal serves as a valuable tool for documenting your personal challenges, moments of growth and the transformations you experience in becoming an integrated STEM teacher.

We encourage you to share your journey with your colleagues and be a mentor to others who wish to start an integrated STEM journey of their own. Lastly, we hope that this is the start of many meaningful connections and collaborations as you continue your integrated STEM learning journey.

HOW TO USE YOUR STEM JOURNAL

This STEM learning journal showcases **six integrated STEM activities** for grades 3 to 12, each presented through the lenses of a different teacher, as they enact and reflect on their integrated STEM teaching experience. The activities are paired with features that use processes such as questioning, reflecting, and journalling to build upon existing ideas or synthesise new ones to facilitate integrated STEM inquiry in your classroom.

Key Features	Purpose
Probe Activities	Probe activities each present a problem, issue, or scenario. They introduce students to complex, persistent and extended problems in a real-world context.
Images of Integrated STEM Inquiry	Each chapter features a vignette of an actual lesson enactment of an integrated STEM activity. These images are likened to videos of lesson enactments and serve to help you reflect on how you can adopt or adapt the enactments for your lessons.
Instructional Model	The instructional models present a summarised flow of the enacted lesson.
Instructional Strategies	Instructional strategies present suggestions for adapting the STEM activity featured in the vignette to suit the learning experiences and profiles of different grade levels.
Reflection Questions	Reflection questions help you reflect on potential challenges that may arise and foresee areas that warrant further development or modification, to help you develop greater STEM teaching self-efficacy.
Journalling Space	Space is provided for you to pen your thoughts, draft lesson ideas, and reflect on your lesson enactment. It also serves to document your personal growth as a STEM teacher.
QR codes	Scan the QR code to access short video clips or editable teaching resources such as lesson plans and the student workbook.

ANCHORING INTEGRATED STEM INQUIRY IN YOUR CLASSROOM

Integrated STEM inquiry focuses on the **problem-solving** process. It involves understanding the problem presented, generating a possible solution, and evaluating, testing, and refining the solution.

The questions asked and the investigations carried out thereafter are important aspects of the STEM inquiry process.

In this book, we use the **STEM Quartet instructional framework** to design integrated STEM activities that begin with a single lead discipline. Emphasis is placed on how the knowledge and skills of the lead discipline are connected to the other disciplines of STEM.

Roger Bybee's 5E STEM framework (Engage, Explore, Explain, Elaborate and Evaluate) is then applied to reflect on the lesson enactment.

At each phase of the 5E framework, we consider:

(1) what the teacher does

(2) what the students do, and

(3) what evidence of students' learning (of STEM) are present.

In particular, we note 10 essential characteristics of STEM inquiry in the classroom.

Scan this QR code to find out more about the STEM Quartet Instructional Framework.

What is the STEM Quartet? Use this space to note some salient points.

Engagement (Problematising)	1.	A problem is presented in a real-life context.
	2.	Guidance is given to separate the problem from the context.
Exploration (Science Inquiry)	3.	Scientifically-oriented questions are used to engage students.
	4.	Students are encouraged to ask scientifically-oriented questions.
Explanation (Group Problem Solving & Design)	5.	Opportunities for discussion and group problem solving are given.
	6.	Students prioritise the use of evidence to develop, evaluate and design a solution to scientifically-oriented questions.
	7.	Students engage in the iterative design process to evaluate and modify their solution based on alternative scientific solutions.
Elaboration (Making Connections)	8.	Guidance is given to foster meaningful connections between the STEM disciplines and solving the problem.
Evaluation (Communication & Feedback)	9.	Students communicate and justify their proposed solution with supporting evidence.
	10.	Peer evaluation and feedback is given.

As you progress through this book, we hope that you can consider how the **characteristics of STEM inquiry** and the **strategies** used to fulfil them may look like in your classroom.

For example, when designing an integrated STEM activity, you may wish to start with a **complex, persistent**, and **extended** problem, and map the students' learning experiences to the intended learning outcomes. Alternatively, you can list the specific learning outcomes of various STEM disciplines and design a problem that requires students to apply the related knowledge and skills to solve it.

Regardless of your design approach, we hope that this book offers you a **personalised STEM learning journey**, streamlined to aid you in designing integrated STEM activities of your own that are tailored uniquely for the context(s) of your classroom.

Before you begin reading the first STEM activity, take a moment to reflect on your STEM teaching experience.

Which of these do you find more challenging — (1) designing a STEM lesson or (2) enacting it? Why?

PLAN YOUR INTEGRATED STEM JOURNEY

Activity One: Urban Farming

Activity Two: Rusting

Activity Three: Oil Spill

Activity Four: Test Kits

Activity Five:
The Neuromuscular System

Activity Six:
The Cardiovascular System

ACTIVITY ONE
URBAN FARMING

What crops would you plant?

Probe Activity

Where Does Our Food Come From?

The probe activity below can be used to tease out and gauge students' prior knowledge on sustainable living. It is designed to initiate discussion and to illuminate students' awareness of the efforts being made to cultivate a greener and more resilient economy.

Scenario

A father and his son were buying groceries at the supermarket. While placing various food products into the trolley, his son Jared noticed that the products came from different countries, such as Vietnam, Indonesia, China and Thailand.

Why do you think only 1% of Singapore's land is used for farming?

Classroom Tidbit
To guide students to describe Singapore as "small" or "an island", you may wish to ask, "Compared to countries such as Australia and Malaysia, how would you describe Singapore's size?" What are some other guiding questions you can use? Note these down.

How would increasing Singapore's food supply be beneficial? In what ways can Singapore accomplish this?

Classroom Tidbit
Have students recall some reasons for food shortages during the COVID-19 pandemic. Then, have students consider, "In what ways can you, as a resident of Singapore, help increase our local food supply?"

Learning Outcomes Matrix

The table below lists some relevant science concepts that can be covered in this integrated STEM activity on food sustainability with urban farming.

Science Concepts/Grade Level	3—6	7—10	11—12
Carbon cycle		√	√
Climate change and global warming			√
Diversity of things		√	√
Environmental protection and conservation	√	√	√
Interactions within ecosystems		√	√
Man's impact on the environment		√	√
Plant system (parts and functions)		√	√
Plant system (photosynthesis)	√	√	√
Pollution	√	√	√
Properties of materials	√	√	√
Renewable and non-renewable resources	√	√	√
Sustainability			√

What other science concepts or learning outcomes would you include in your lesson?

Implementing Integrated STEM Inquiry in the Grade 11—12 Classroom

The vignette on the next page offers insight into how, "Where does our food come from?", can be expanded on and contextualised to facilitate STEM inquiry in a grade 11–12 classroom.

The vignette features an integrated STEM activity designed by a teacher, Ms Yeo, to introduce the concept of sustainability and urban farming in land-scarce countries and cities.

Integrated STEM Activity		
Part	Task	Time (hours)
One	Introduction & Research	1.0
Two	Prototype Design & Construction	2.75
Three	Prototype Presentation & Evaluation	0.75

Find out how Ms Yeo uses the 5E instructional model to integrate STEM knowledge, skills, and practices into her grade 11–12 classroom to teach her students about vertical farming.

Included in the vignette are examples of her students' responses and her reflections on the process. The characteristics of STEM inquiry relevant to her lesson are noted in the margin on the left.

Scan this QR code to watch a video on vertical farming!

Guidance to identify the problem

Ms Yeo begins her class with the probe activity. She guides her students through the questions by first showing a map of Singapore.

She uses this as a visual aid to illustrate Singapore's small size compared to larger countries across the globe, by superimposing it onto a lake in New Zealand.

Presenting a problem in a real-life context

Ms Yeo then prompts her students to consider: (1) how land is used, and (2) the potential challenges of meeting the needs of fast-developing small cities/countries like Singapore.

One student suggested that Singapore's economic growth is in part reliant on tourism; hence, land is required for tourist sites. Another suggested the need of land for housing and industrial production. When a student remarks that land in Singapore is also used for vertical farming, Ms Yeo seizes the opportunity to invite the class to recall what they know about vertical farming.

Teacher-directed scientific inquiry

For example: Where have you seen vertical farming in Singapore? What do you think vertical farming is used for in Singapore?

Guidance to identify the problem

Ms Yeo further engages her students by inviting them to tap into their personal experiences.

For example: Recall things that you observed while travelling in countries larger than Singapore. What types of places did you visit? Were the places vast, urban, rural, used for housing, farming and so on?

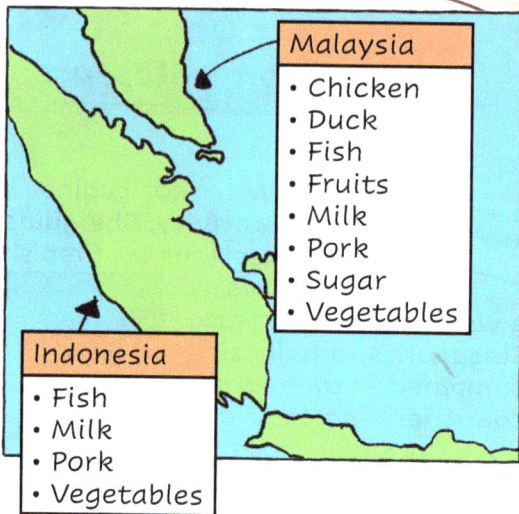

Malaysia
• Chicken
• Duck
• Fish
• Fruits
• Milk
• Pork
• Sugar
• Vegetables

Indonesia
• Fish
• Milk
• Pork
• Vegetables

She builds on the students' responses of "large areas of land" and "farms" by having them discuss different types of agricultural practices used in these larger countries. While many gave examples of animal farming, one student shared about their visit to a palm oil plantation in Malaysia.

This afforded Ms Yeo the opportunity to raise other examples of crop farming, such as terrace farming and the rice paddy fields in Thailand, both of which require large areas of land. She then poses a critical thinking question to help her students engage in more in-depth comparisons and critical thinking.

Teacher-directed scientific inquiry

Critical thinking question: How does Singapore obtain its food given that land is scarce?

Guidance to identify the problem

Her students were quickly reminded of Jared's observations in the probe activity, allowing Ms Yeo to follow up with a chart on the major sources of food for Singapore, to enforce that 90% of the country's food is imported. She then invites her students to consider two overarching questions.

Teacher-directed scientific inquiry

Critical thinking questions:

1. Why would a high import rate not be ideal for small cities or countries?

2. How can small cities and countries still thrive despite land scarcity and limited import of food, such as during a time of crisis?

How would you present the concepts of land scarcity and sustainability in your classroom?

Aware that her students may not intuitively associate the positive socioeconomic impacts of urban farming practices, such as vertical farming, in augmenting food sustainability in land-scarce areas, she scaffolds their critical thinking through teacher-directed inquiry.

For example, she breaks down the first overarching question into simpler questions:

Guidance to identify the problem

Making connections

Teacher-directed scientific inquiry

1. What does the term "import" mean?

2. What does it mean when a city or country has a high import rate?

3. What would happen if there was a disease outbreak in this city and its borders were to close?

4. Where would the city obtain its food from?

This helps her students build a more tangible understanding of the problem of land scarcity and the benefits of vertical farming before they attempt an integrated STEM challenge, set in the context of Singapore's 2030 Green Plan agenda for sustainable development.

Integrated STEM Activity

You have been tasked to design and build a prototype of a vertical farm for your school community garden. The farm must improve the efficiency of the community garden and increase your school's food production in a sustainable way. There are three criteria your design must adhere to: (a) limited availability of land (approximately half a hectare of land), (b) limited manpower and (c) cost-effectiveness.

What other constraints or limitations might you include?

Group problem solving

Student-centric inquiry

To facilitate the research and discussion processes, the students work in groups of four or five and each member is assigned one of three roles:

(1) to read up and evaluate information provided by their teacher;

(2) to conduct an online search for additional information on vertical farming; or

(3) to note down the salient points.

The students are then invited to share what they have learnt with the class. As students were not restricted in their information search on vertical farming, a broad spectrum of information is shared. One student shares that an advantage of vertical farming is that "more crop [can be grown] per square footage of area", while another student notes three methods of vertical farming — hydroponics, aeroponics and aquaponics. Through this short exercise, Ms Yeo notes that self-directed learning offers students an opportunity to pursue different lines of inquiry.

To continue their discussion, Ms Yeo has her students recall the important factors for "good plant growth", drawing on their prior knowledge on photosynthesis to highlight other factors that the students should consider in their prototype design.

She ends the first part of the integrated STEM lesson by emphasising that a:

"good prototype design should provide a solution that is

feasible

practical

efficient

user-centric, and

user-friendly."

Which aspect(s) of part one would you adopt and why?

Part Two:
Prototype Design & Construction

Ms Yeo starts the ideation process by first having her students work individually to jot down or sketch as many different "big picture" ideas for their vertical farms as they can. She tries to create a safe space for her students by reassuring them that there are no right or wrong solutions. In addition, a 15-minute time frame is posed to kickstart the brainstorming process and expedite idea generation.

Communication

Group problem solving

Peer evaluation

The students then work in groups to share their thoughts, using their sketches or notes as points of discussion. When reminded to factor the constraints discussed into their designs, most groups noticeably engage in more in-depth discussions and can be seen comparing and evaluating different sketches, providing peer feedback.

While the students gather and sketch their design ideas, Ms Yeo encourages them to include as much detail as possible. She explains that the diagrams act as tools to communicate ideas visually, in a quick and simple way.

Prioritising evidence

Group problem solving

The detailed diagrams also allow her to instigate scientific discourse by having the students justify their design choices using scientific evidence and language. As she makes her rounds, Ms. Yeo is pleased to see evidence of collaboration. While some groups fuse features from the designs of more than one group member, others work together to modify and improve a selected design.

Teacher-directed scientific inquiry

Ms Yeo also challenges her students to reflect on how balance between efficiency through automation and cost-effectiveness can be achieved.

For example:

Are there ways you can make your farm semi-automated without using smart technology? What role could recyclable materials play in the construction of your vertical farm?

Ms Yeo then takes a step back and supervises them with minimal guidance to foster self-agency, peer-to-peer learning and collaboration.

Which aspect(s) of part two would you adopt and why?

What are some difficulties that you foresee your students having in the design and construction process?

Part Three:
Prototype Presentation & Evaluation

Communication

In lieu of time, Ms Yeo starts the third part of the session by pairing groups to evaluate each other. To facilitate the review process, she reiterates the three key constraints, advising students to look for "blind spots" — factors that have not been addressed.

All materials used are recycled! Plants that need more sunlight are closer to the top.

This system is inside a pyramid structure and uses rainwater collected in a tank and water and nutrients gathered from planter holes.

Each group:

1. presents a sketch of their prototype,

2. explains how it works,

3. describes how the constraints have been addressed, and

4. justifies how their design solves the problem.

Peer evaluation

Group problem solving

Feedback is then given by the partner group. Following evaluation, the students review and discuss the shortcomings of their solution(s) based on the feedback.

Engaging in iterative design

At this stage, students enter an **iterative process** of (re)design and (re)evaluation, as they collaborate to find and deliberate on alternative solutions or make improvements to modify their design.

Ms Yeo noted that this process provided an opportunity for students to further build on their knowledge, using critical thinking and analytical skills, as they source for and evaluate more information to support their design changes.

Communication

Peer evaluation

The students then undergo a second paired peer review session, presenting their improved design solutions with explanations on how the feedback was incorporated. This is followed by a final session for further research and revisions to their prototypes' designs before beginning the construction process.

Group problem solving

Students have access to various materials, such as straws of different sizes and colours, cardboard, ice cream sticks, plastic containers, and plastic mesh to build their prototype. Ms Yeo urges them to think creatively and innovatively and encourages them to be less literal with the materials provided.

For example: How can simple materials be used to represent more complex things or systems? Could these straws be used as something other than pipelines? How about this cardboard, can it only serve as structural material?

Communication

Peer evaluation

Teacher evaluation

Making connections

Lastly, each group of students gives a final presentation to their classmates and teacher. Ms Yeo creates an open environment for discussion by welcoming questions and comments from the students before sharing her own. She ensures to highlight one or more positive aspects for each prototype. At times, Ms Yeo will elaborate on how the prototype could work in a real-world context, helping students to make a deeper connection with the purpose and significance of the integrated STEM activity.

Scan this QR code to access a suggested list of materials for this activity and how they can be used.

Which aspect(s) of part three would you adopt and why?

Classroom Tidbit
Some students may find it difficult to "think outside the box" while using everyday materials. You may wish to note down some suggestions in the space provided for various materials.

Overall, which aspects of Ms Yeo's lesson would you find most difficult to replicate in your classroom?

What are some ways you can assess your students' learning in this activity?

What type of vertical farm would you design given the same constraints? Sketch your prototype in the space below!

Scan this QR code to access the student workbook for this activity.

Ms Yeo's Instructional Model

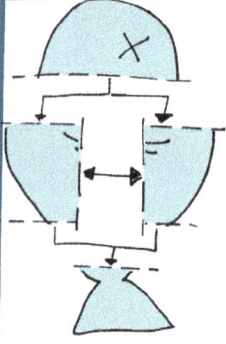

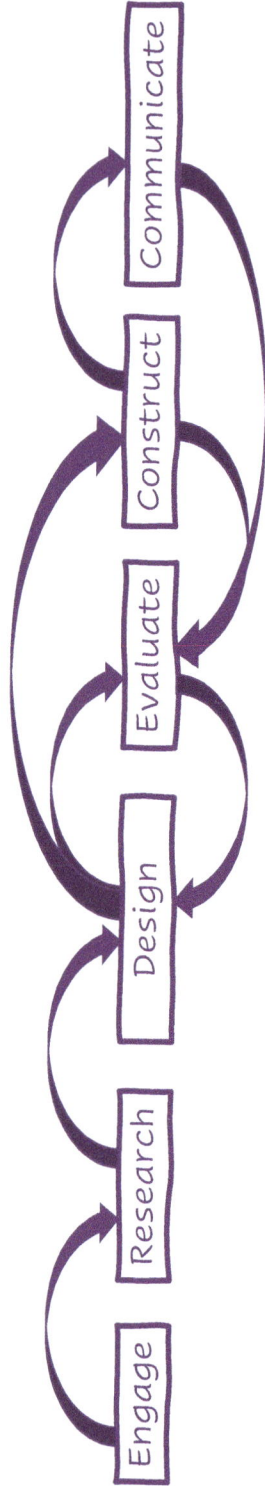

Engage → Research → Design → Evaluate → Construct → Communicate

Scan this QR code to access the detailed and editable instructional model.

Instructional Strategies

Students will:

Grade 11–12

Back to page 6!

- Understand the impacts of climate change and emerging technologies on the environment
- Understand Man's role and responsibility in conservation and creating a sustainable environment
- Learn about different types of urban farming and their benefits and disadvantages
- Design and build a school community garden that is limited in land, cost-effective and limited in manpower

Classroom Tidbit

Unsure of how to adapt Ms Yeo's vertical farming activity for students of different grade levels? Pick your grade level, then follow the instructions.

Grade 9–10

Skip to page 25!

- Understand how different systems interact and affect each other
- Understand the factors that can affect an ecosystem and population growth
- Understand the impact of climate change and emerging technologies on the environment
- Learn about different types of vertical farming, their benefits and disadvantages
- Design and build a school community garden that is eco-friendly and promotes biodiversity and nutrient cycling, while being limited in land and being cost-effective

Grade 7–8

Page 24 it is!

- Understand Man's impact on the Earth's renewable and non-renewable resources
- Learn about vertical farming and its benefits in a land-scarce city or country
- Design and build a school community garden that improves food availability and sustainability, while being limited in land and being cost-effective

Grade 5–6

Off to page 23!

- Understand Man's impact on the Earth's water resources
- Learn about vertical farming and its benefits in a land-scarce city or country
- Design and build a school community garden that is limited in land and cost-effective

Grade 3–4

Flip the page!

- Recall the characteristics of living things, in particular, what plants need to grow and survive
- Learn about vertical farming
- Design and/or build a solution to overcome the lack of land on an increasingly populated island for farming

For the grade 3–4 classroom

At this level, students learn broadly about diversity. They are introduced to living things. They learn that plants are a source of food for man, and need air, water, sunlight and nutrients to grow. The probe activity can be used to guide students to understand the concept of land scarcity. A hypothetical scenario in place of the school community garden challenge can be given, to stimulate discussion and exploration of unconventional farming solutions.

Scenario: Imagine that you live on a very small island. There is no transport to and from this island. Residents can only grow and obtain their food from a single plot of land that they share. As time passes, the population continues to grow and more land is needed for housing. With lesser availability of land and no access to other food sources, what can you build or design to make sure the residents have enough crops to eat?

Students should not be expected to define sustainability. However, they can be encouraged to wonder: (1) how a smaller plot of land can accommodate more crops, and (2) how existing unconventional farming areas can be used for farming. Vertical farming can then be introduced without covering the concept of urban farming and the specific details of various types of vertical farms.

What would you adopt or adapt from this activity to suit the learner profiles in your classroom?

At this level, students learn to be responsible for plants and animals, and touch on water pollution and the importance of conserving water. The probe activity can help students reflect on why land is also an important resource, particularly in small countries/cities. Students can then be shown various pictures of traditional and urban farming practices and be guided to notice that traditional farming typically requires more land than urban farming. Students can thereafter explore various types of urban farming and be guided to understand how vertical farming is suitable and sustainable for Singapore to increase local food production. Students can then complete a simplified version of the grade 11–12 integrated STEM challenge, as students may find the concept of automation difficult to grasp.

Task: You have been tasked to design and build a prototype of a vertical farm for your school community garden. How can you design a farm that improves the efficiency of the garden and increases its food production in a sustainable way?

Lastly, the students can be challenged with two criteria: (a) limited availability of land (approximately half a hectare of land) and (b) cost-effectiveness.

What would you adopt or adapt from this activity to suit the learner profiles in your classroom?

In grades 7–8, students gain a deeper appreciation of the Earth's resources (renewable and non-renewable) and how living things depend on them for survival. They become aware of ways to use the resources responsibly and sustainably, and how to conserve them. Following the probe activity, students can be guided to think critically on how population growth and urbanisation play a role in land scarcity, availability of local crop, pollution, and loss of biodiversity. Students can then be introduced to urban farming as a sustainable practice in urban cities and countries. Students can be challenged with a modified version of the grade 11–12 integrated STEM activity.

Task: You have been tasked to design and build a prototype of a vertical farm for your school community garden. Aside from urban farming, what other practices can help your school improve crop availability sustainably? Include one or more of these into the design of your vertical farms.

Students can be given two criteria: (a) limited availability of land and (b) cost-effectiveness. Students can be challenged to incorporate ways to minimise food waste and water consumption into their designs.

What would you adopt or adapt from this activity to suit the learner profiles in your classroom?

By grade 10, students recognise how Man's interactions with the environment can affect the ecosystem and cause change. For example, they recognise how abiotic factors (e.g., air, water, temperature, etc.) can affect the environment and the survival of organisms. They can also appreciate how climate change can influence how we interact with the environment and the resultant technologies that emerge. At this stage, students have also learnt about food webs, the non-cyclical flow of energy, and nutrient recycling. Students can thus be presented with a modified version of the grade 11–12 integrated activity.

Task: You have been tasked to design and build a prototype of a vertical farm for your school community garden. What are some ways you can design an eco-friendly vertical farm that promotes urban biodiversity and nutrient cycling?

Students can first explore different types of vertical farming and consider their advantages and disadvantages. They can then find ways to overcome some of the disadvantages like high-power consumption and pest infestation.

What would you adopt or adapt from this activity to suit the learner profiles in your classroom?

Reflection Questions

Now that you have been through this integrated STEM activity, do you have a better idea of what a STEM lesson looks like?

How confident are you to carry out this activity in your classroom? Colour the fish to represent your confidence level and note down some reasons why.

Scan this QR code to access Ms Yeo's editable lesson plan.

ACTIVITY TWO
RUSTING

How can rust be prevented?

Probe Activity
Why Do Things Rust?

This probe activity introduces rusting in an everyday context. It is designed to elicit students' knowledge on chemical change and oxidation, particularly in a climate/condition where rusting is accelerated.

Scenario

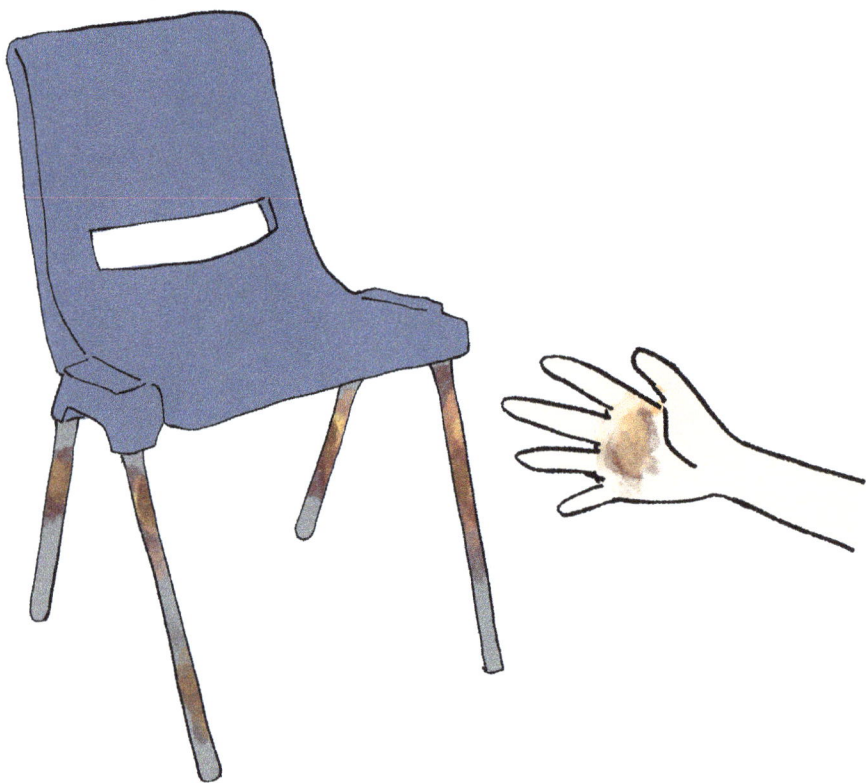

At the end of every school day, the chairs in each classroom are stacked on top of each other and the tables are placed against the wall. Today, Mr Francis has asked his students, Nam and Tony, to carry out this task. After stacking the chairs, Tony notices brown-orange stains on his palm. He wipes his palm against his uniform to remove them.

Which of the two students is correct? Are dust and rust different?

The table below lists some relevant science concepts that can be covered in this integrated STEM activity on rusting and its damaging effects on infrastructure.

Science Concepts/Grade Level	3—6	7—10	11—12
Atoms, molecules and compounds		√	√
Balancing chemical equations		√	√
Chemical (and physical) changes	√	√	√
Composition of matter	√	√	√
Conditions for corrosion (rusting)		√	√
Electroplating (electrophoresis)		√	√
Hydration reactions			√
Ionisation (gain and loss of electrons)		√	√
Ionisation energy			√
Ionic radius			√
Methods of corrosion prevention		√	√
Reactivity series of metals (sacrificial metals)		√	√
Redox reactions with metals		√	√

What other science concepts or learning outcomes would you include in your lesson?

Implementing Integrated STEM Inquiry in the Grade 7—8 Classroom

Vignette

The vignette on the next page offers insight into how "Why do things rust?" can prompt STEM inquiry on corrosion and infrastructure degradation.

The vignette describes how Ms Kee adopts the S-T-E-M Quartet instructional framework to introduce STEM knowledge, skills, and practices into her grade 7—8 classroom for the first time.

Integrated STEM Activity		
Part	Task	Time (hours)
One	Problem Identification	
Two	Deciding on the Nature of the Problem	1.0
Three	Planning & Proposing a Solution	

The featured integrated STEM activity is problem-centric and requires students to draw on and develop skills from chemistry (lead discipline) and engineering to design a solution that reduces or prevents rusting on a bridge in an area of high humidity. Included in the description are examples of her students' responses and her reflections on the process. The characteristics of STEM inquiry relevant to her lesson are noted in the margin on the left.

Scan this QR code to watch a video on rusting.

Past experience has shown Ms Kee that students are not always able to fully appreciate the negative impact of rust and the importance of understanding its underlying chemistry. Ms Kee has hence adopted the S-T-E-M Quartet Instructional Framework to design a STEM activity that would help students foster meaningful connections between the complex, persistent and extended problem of rust and the relevant STEM disciplines from which content knowledge and skills are required to prevent it. The activity is centred around the real-world problem of rusting and its adverse effects on infrastructure.

To begin her lesson, Ms Kee asks her students to consider the probe activity, "Why do things rust?" She uses a series of scientifically-oriented questions to prompt students to analyse the conversation between Tony and Nam and think critically to determine which student is correct.

Teacher-directed scientific inquiry

For example, "What are some keywords or observations you can use to help you answer the questions?" and "How are dust and rust similar or different in appearance?"

She encourages her students to describe the physical properties of dust and rust before inviting them to recall where they have observed them. This enables her to guide her students to identify a common theme — that rusting occurs on (some types of) metals, specifically, metals that contain iron.

Making connections

Next, she draws on their prior knowledge of physical and chemical changes, by asking them to recall each definition. She scaffolds the critical thinking process again with a series of scientifically-oriented questions to help them associate dust with physical change and rust with chemical change. Ms Kee then deliberately links the concept of rusting and redox reactions by showing her students how to derive the chemical equation for iron oxide.

What are some simple demonstrations you can carry out or prepare for this topic? Note these down in the space provided.

Classroom Tidbit
For younger grade students or students that are learning about rust for the first time, try conducting a demonstration to show how dust and rust are formed.

Scan this QR code to find out how you can make rust fast!

Classroom Tidbit
If your students are learning about physical and chemical changes for the first time you may wish to demonstrate why an action or event is classified as a physical or chemical change. An editable worksheet has been provided for added practice.

Scan this QR code to access a worksheet on physical and chemical changes!

Presenting a problem in a real-life context

Once her students have understood that rusting is a type of redox reaction, she presents them with a task.

Integrated STEM Activity

You have been hired, as a member of Rust Busters to design a bridge that connects a small island off the coast of Singapore to its mainland.

How can you design a bridge that does not rust easily under humid conditions?

As you design your bridge, you must consider:

- The conditions required for rust formation
- How rusting is accelerated in a hot and humid country (such as Singapore)
- Methods of rust prevention or reduction
- How the rate of rusting will differ in other climatic zones

Students are then given worksheets and access to the online learning and sharing platform, Nearpod. Students begin by embarking on an individual learning journey using a ready-to-run interactive lesson designed by Ms Kee on Nearpod.

The lesson consists of learning slides, links to resources, such as videos, and quiz questions. The online lesson was designed to engage students in self-paced learning with technology as a learning tool.

Content knowledge on rusting and methods of rust prevention is scaffolded. This is reflected in the quiz questions used to assess their learning throughout the online lesson. The questions increase in difficulty as students progress through the lesson and build on their content knowledge.

Would an online platform such as Nearpod be an effective way of encouraging self-paced learning in your classroom? Why?

What other ways or platforms could you use to engage your students in self-paced learning?

Guidance to identify the problem

Ms Kee uses Nearpod's real-time analysis function to review her students' quiz performance as they progress through the online lesson. This enables her to quickly identify and address any gaps or misconceptions in their newly acquired knowledge before moving on. The students then watch a video of an experiment on the conditions required for rusting to occur. The aim of the video is to help students identify how and why rusting occurs. It also acts as a segue for Ms Kee to invite her students to brainstorm potential situations and problems that can arise with rusting.

| Only nails | Nails in rice grains | Nails in salt water | Nails in oil | Nails in water |

Making connections

Next, she follows up with two case study videos featuring the Golden Gate Bridge and Lowe's Motor Speedway in America, and a virtual scientific investigation on the conditions for rusting. These videos help reinforce students' understanding of how rusting can cause structural change. Importantly, they illuminate how knowledge from various disciplines (within and outside) of STEM contribute to understanding the nature and causes of rusting, its sociocultural impacts and the potential solutions that can resolve its problems.

How would you assess the gaps in your students' content knowledge differently from Ms Kee?

What potential misconceptions do you anticipate your students having?

Classroom Tidbit
To gauge the depth of your students' content knowledge on the topic, you may wish to ask, "Are rusting and corrosion the same?" or "Can all metals rust?"

What alternative case studies could you use to help your students establish a deeper connection with the problem of rusting?

Classroom Tidbit
Search for potential case studies that are closer to home. Note some of these in the space provided.

Group problem solving

Once Ms Kee is satisfied that her students have grasped the content knowledge and problem at hand, she divides them into groups of four. Students then engage in group work to research, brainstorm and problem-solve potential solutions to prevent or reduce rusting on a bridge in an area of high humidity.

Communication

Prioritising evidence

While exploring information on potential solutions, students come across various methods of rust prevention, such as paint-coating, oiling/greasing, electroplating, using sacrificial metals and galvanising. Ms Kee encourages them to share their ideas openly and consider how the different methods of prevention may or may not work in Singapore's climate. Importantly, she reminds them to support their arguments with appropriate scientific evidence as they draft their final ideas and solutions and upload them onto Nearpod.

Teacher evaluation

As Ms Kee makes her rounds, she observes the ideas and discussions from each group, noting down the challenges and learning problems they encounter. As a whole, she notices that many of the students grapple with integrating and relating the knowledge and skills from the different disciplines (notably geology, chemistry, and engineering) to their proposed solution.

Teacher evaluation

In particular, Ms Kee notes that the groups exhibit common challenges. Firstly, they experience idea scarcity and difficulties in crafting creative and innovative solutions, with many students opting to use stainless steel and/or paint-coating to prevent rusting. A group of students even refer to their solution as "basically any bridge".

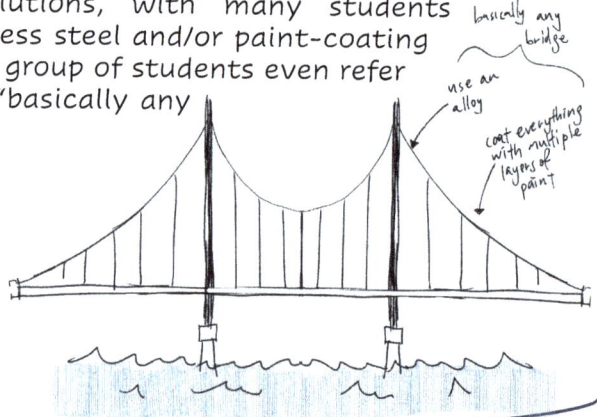

basically any bridge

use an alloy

coat everything with multiple layers of paint

What challenges do you foresee your students having? How can you help them overcome these challenges?

What are some ways you can inspire your students to prevent idea scarcity?

Classroom Tidbit
Show unique designs of bridges from different countries and challenge your students to describe how a particular design can influence its function. Alternatively, you may wish to draw on their prior knowledge on materials and have them consider alternative materials to iron and/or steel.

In contrast, there was one student whose idea to build a large "umbrella to shield rain?" diverged from the status quo. However, as her design was not supported by existing ideas on the internet, she included a "question mark" with her comment.

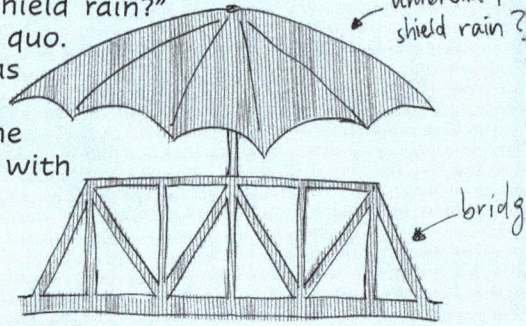

umbrella to shield rain?

bridg

Teacher evaluation Ms Kee postulates that the question mark could represent self-doubt, unease or uncertainty when expressing unconventional ideas.

She reasons that the students' struggle to engage in creative and innovative thinking may be the result of the short one-hour time constraint — that their apprehension to explore and express ideas that fall "outside the box" may be due to a need for more exposure to the integration of STEM disciplinary knowledge and practices. As this is the first time her students are encountering the engineering design process and the concept of rusting, they may not possess sufficient knowledge to address the multifaceted issues associated with rusting and infrastructure.

Teacher evaluation Secondly, she notices that the students tend to be overreliant on the internet, choosing to use easily accessible technology to facilitate idea generation instead of actively engaging in the problem-solving process. Upon reflection, Ms Kee recognises the importance of making time and room for her students to consider unconventional ideas or solutions, and that emphasis should be placed on the infinite number of potential solutions that can arise from a complex problem, rather than just a single solution.

Teacher evaluation Ms Kee proposes that teachers can encourage students to explore new lines of scientific and engineering-based inquiry by scaffolding their online search with suggested keywords to bring new or different perspectives to light. This could assist students in nurturing their critical thinking skills and attention to detail as they learn from exisiting ideas and build on their knowledge base. In addition, it could also increase confidence and promote ideation fluency.

More time and guidance are also needed for students to not only think creatively and innovatively, but to also conduct more in-depth research and evaluation of information, group discussion and problem solving, and engage in the iterative process of prototype design and testing.

How can you encourage your students to explore other sources of information and reduce their reliance on the internet? What incentives could you give to facilitate this process?

What are some keywords you could suggest to scaffold your students' learning and information search?

Classroom Tidbit
Try drafting a flowchart to illustrate the potential lines of inquiry that your students could pursue for each keyword.

Teacher evaluation

The third common challenge Ms Kee observed is the inaccurate or incomplete graphical representations of their predictions.

The majority of students had:

1. incorrect or incomplete labelling of axes and/or

2. incorrect placing of the dependent and independent variables on the X and Y axes.

This highlights that further development of students' content knowledge on rusting and the standard practices of graphing are still required for students to adapt more seamlessly to an integrated learning framework.

Teacher evaluation

Ms Kee emphasises that support should be implemented early on in the lesson. Such support can range from scaffolding new ideas and expectations, to providing a review section of the pre-requisite knowledge from the relevant disciplines, to illuminate the connections between the disciplinary content, ideas, practices and the real-world problem.

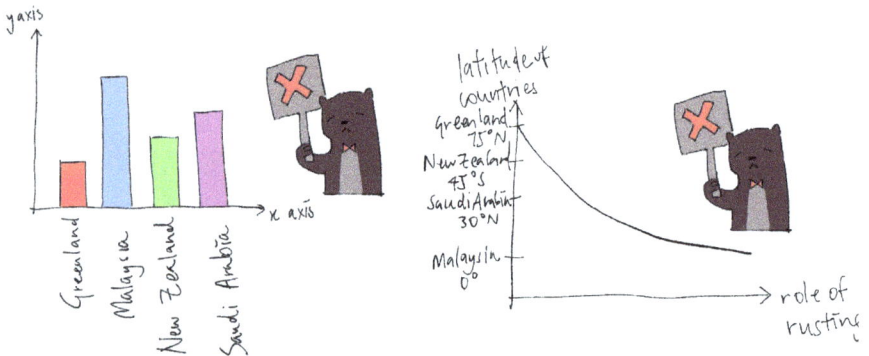

Teacher evaluation

In her final reflection, Ms Kee also notes that some key features of STEM inquiry in the classroom have not been addressed. Although the students designed, illustrated and explained their solutions, they did not build or test a prototype of their design. Ms Kee is aware that the engineering design process also involves components of model-making and repeated cycling between research, design, evaluation, testing and refinement. However, she reasons that students are often given limited time to complete class activities.

In order for students to benefit from a more holistic and integrated STEM learning experience, more classroom time should be apportioned towards the engineering design process and overall instructional time.

Which aspect(s) from Ms Kee's lesson would you adopt or adapt, and why?

Which aspect(s) from Ms Kee's lesson would you change, and why?

Ms Kee conducted her STEM activity in one hour. Would this time frame be suitable for your students? How much time would you allocate to each part of the lesson and why?

Scan this QR code to access the student workbook for this activity.

Ms Kee's Instructional Model

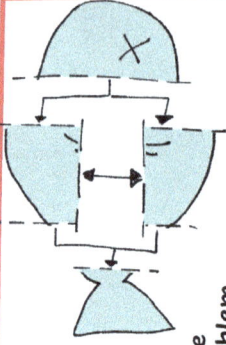

2. Deciding on the nature of the problem

Mathematics

Engineering

Problem Scenario

Technology

Science

1. Problem identification

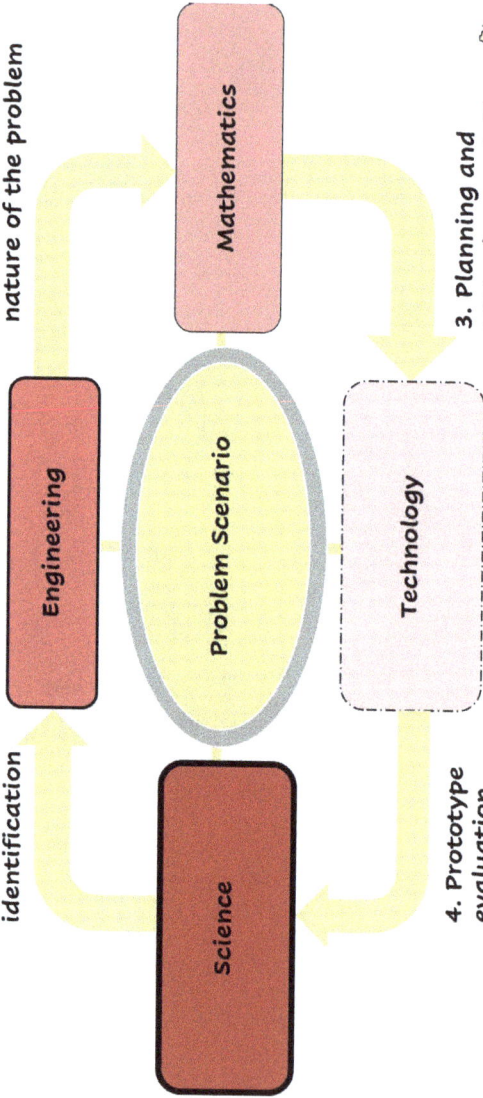

3. Planning and proposing a solution

4. Prototype evaluation

Scan this QR code to access the detailed and editable instructional model.

Instructional Strategies

Students will:

Grade 11–12

Page 49 it is!

- Recall conditions for rusting
- Recall the redox reaction for rusting
- Recall methods of rust prevention
- Design an experiment to compare three methods of rust prevention (electroplating, sacrificial protection and galvanisation)
- Design a bridge that does not rust easily under Singapore's varying weather conditions

Classroom Tidbit
Unsure of how to adapt Ms Kee's rusting activity for students of different grade levels? Pick your grade level, then follow the instructions.

Grade 9–10

Skip to page 48!

- Recall conditions for rusting
- Recall the redox reaction for rusting
- Understand methods of rust prevention
- Design an experiment to explore the impact of varying weather conditions in Singapore on rusting
- Design a bridge that does not rust easily under Singapore's varying weather conditions
- Explore other metals/alloys as a potential substitute for iron

Grade 7–8

Back to page 31!

- Recall properties of metals and alloys
- Recall and understand examples of physical and chemical changes
- Learn about rusting and the conditions required for it to occur
- Explore methods of rust prevention
- Design and build a bridge that does not rust easily under Singapore's humid weather
- Predict how the rate of rusting will change under different climatic conditions

Grade 5–6

Off to page 47!

- Learn about metals, alloys and their properties
- Learn about physical and chemical changes
- Learn about rusting and the conditions required for it to occur
- Design an experiment to find out whether an iron alloy rusts at the same rate as its purer constituent
- Understand the concept of a fair test

Grade 3–4

Flip the page!

- Learn about different types of materials and their properties
- Understand some properties of metals and their uses
- Learn about rusting and the conditions required for it to occur
- Explore ways to prevent rusting

For the grade 3–4 classroom

At the grade 3–4 level, students learn broadly about materials and their physical properties. They learn that some metals are strong and not easily broken while others are malleable. The probe in this chapter can be used to introduce one example of how metal is used in everyday life. Students can be guided to understand that iron is commonly used to make steel, a frequently used structural metal in infrastructure. Students can then be invited to observe a shiny nail and a rusty nail and describe what they see. Students can be encouraged to assess the texture of both nails and be prompted to consider:

Where have you seen rust before? What would happen if the objects around you were made of a material that can rust?

Students can learn about the effects and impact of corrosion and the optimal conditions for rusting. They can then be challenged with a modified version of the grade 7–8 integrated STEM activity.

Task: Nails are important tools that help hold structures together. These structures can be as small as a stool or as large as a bridge. You have been hired, as a member of Rust Busters, to find out, "How can you prevent nails from rusting?"

To complete this task, students can brainstorm and research ways to prevent rusting (without having direct knowledge of barrier and sacrificial methods).

What would you adopt or adapt from this activity to suit the learner profiles in your classroom?

At the grade 5–6 level, students learn about alloys and can be invited to compare the properties of some alloys to their purer constituents. Students also learn about physical and chemical changes. The probe can be used to assess what they have learned. Students can then be introduced to rusting as a form of corrosion and the conditions required for it to occur, by watching a video, like the "Nail rusting experiment" in the vignette. They can then be asked to complete a modified version of the grade 7–8 integrated STEM activity.

Task: Nails are important tools that can hold some structures together. These structures can be as small as a stool or as large as a bridge. You have been hired, as a member of Rust Busters, to find out if an iron alloy would rust at the same rate as its purer constituent given the right conditions. Design and conduct an experiment to test this. Then, find out how you can prevent the nails from rusting.

The task introduces students to the concept of a fair test. Students can explore the effects and impacts of rusting on infrastructure, and how to prevent them. At this stage, students do not need to understand the specific processes involved in prevention methods like electroplating and sacrificial metals.

What would you adopt or adapt from this activity to suit the learner profiles in your classroom?

At this level students have learned to balance chemical equations and are aware of various types of chemical changes (e.g., thermal decomposition, combustion and oxidation). The probe can prompt students to recall the conditions required for rusting and the redox reaction that occurs. Students can then be given a modified version of the grade 7–8 integrated STEM activity.

Task: You have been hired, as a member of Rust Busters, to design a bridge that connects a small island off the coast of Singapore to its mainland. Find out, "How can you design a bridge that does not rust easily under the varying weather conditions in Singapore?"

Students can design and conduct an experiment to test the rate of rusting of iron/iron alloys against various conditions (e.g., different temperatures, and the presence and absence of salt water). Students can then be further challenged to compare the effects of corrosion on iron and aluminum and consider: (1) How is corrosion different in iron and aluminum?, (2) Is aluminum a more suitable structural metal? and (3) What other suitable metals/alloys can be used as a substitute for iron-based structural metal?

Students can be guided to understand that unlike iron, aluminum does not rust. Instead, a layer of aluminum oxide forms during corrosion to protect the metal from the environment. In this challenge, students can also be tasked to evaluate the physical and chemical benefits of various metals/alloys versus their cost as structural materials.

What would you adopt or adapt from this activity to suit the learner profiles in your classroom?

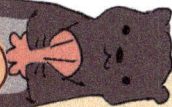

By grade 11–12, students can explain how methods such as electroplating and sacrificial protection prevent rust formation. Galvanisation is a technique that combines both electroplating and sacrificial protection. Students can be given a modified version of the grade 7–8 integrated STEM activity.

Task: You have been hired, as a member of Rust Busters, to design a bridge that connects a small island off the coast of Singapore to its mainland. How can you design a bridge that does not rust easily under the varying weather conditions in Singapore? Find out, "Which rust prevention method (electroplating, sacrificial protection or galvanisation) is most suitable for a bridge built in deep sea water?" Design an experiment to compare and test all three techniques. Support your answer with scientific evidence.

Students can build identical prototypes of their bridge and test the effectiveness of each prevention method. This investigation requires students to develop specific criteria or ways of assessing the effectiveness of each rust prevention technique. At this stage, students should also be able to make other calculations, such as to determine the mass/volume of substances released during electrophoresis and describe the redox processes in sacrificial protection (in terms of ionisation energy and electron transfer, and changes in oxidation state).

What would you adopt or adapt from this activity to suit the learner profiles in your classroom?

Reflection Questions

You have now read two examples of how STEM activities can be implemented in the classroom. In each example, the teachers have adopted different instructional models. From your teaching experience, what instructional model would best facilitate STEM inquiry in your classroom?

How confident are you to carry out this activity in your classroom and why? Colour the fish to represent your confidence level and note down some reasons why.

Scan this QR code to access Ms Kee's editable lesson plan.

ACTIVITY THREE
OIL SPILL

Are all oil spills easy to clean?

Probe Activity
How Do We Clean Up an Oil Spill?

This probe activity below can be used to assess students' prior knowledge on mixtures and immiscibility. It is designed to spark a conversation that underscores the importance of understanding the properties of different materials and substances.

Scenario

Mateo works at a fast-food restaurant. As he was cleaning the deep fryer one evening, he accidentally spills some used oil onto the ground. He quickly picks up a mop and a bucket of water to clean up the spill. However, despite his efforts, he is unable to mop up all the oil. Instead, the oil seems to spread further across the floor each time he adds more water.

What can Mateo do to clean up the oil spill more easily? Why would your method be more successful in cleaning up the spill?

Can you think of some reasons why an oil spill should be quickly cleaned up?

The table below lists some relevant science concepts that can be covered in this integrated STEM activity on understanding the properties of substances and materials and their role in cleaning up oil spills at sea.

Science Concepts/Grade Level	3—6	7—10	11—12
Biomimicry			√
Diversity of materials	√	√	√
Energy sources and their uses	√	√	√
Flow of matter and energy	√	√	√
Hydrocarbon combustion of fossil fuels			√
Hydrophobicity and hydrophilicity (polarity and dipole interactions)		√	√
Interaction (Man and the environment (pollution) and ecosytems)	√	√	√
Oleophobicity (lacks affinity for oils) and oleophilicity (strong affinity for oils)		√	√
Mixtures and immiscibility	√	√	√
Physical and chemical properties of fossil fuels		√	√
Separation techniques (filtration, evaporation, burning)		√	√
Surface tension and density		√	√

What other science concepts or learning outcomes would you include in your lesson?

Implementing Integrated STEM Inquiry in the Grade 5—6 Classroom

The vignette on the next page offers an insight into how, "How do we clean up an oil spill?" can be utilised to design a STEM activity that explores: (1) sources of oil (energy), (2) its uses, (3) the hazards of extracting, processing and transporting oil on the environment, and (4) the methods and tradeoffs for cleaning up an oil spill.

Integrated STEM Activity		
Part	**Task**	**Time (minutes)**
One	Ask	60
Two	Research, Imagine & Plan	60
Three	Create, Test & Improve	60
Four	Communicate	60

The featured integrated STEM activity delves into how Mr Williams adopts the engineering design process to encourage his students to explore and appreciate the chemistry of substances and materials.

Included in the description are examples of his students' responses and his reflections on the process. The characteristics of STEM inquiry relevant to his lesson are noted in the margin on the left.

Scan this QR code to watch a video of an oil spill at sea!

Part One: Ask

Presenting a problem in a real-life context

Teacher-directed scientific inquiry

Mr Williams prepares the following materials before his lesson: a glass tray, a small cup of used cooking oil, a mini mophead fashioned out of yarn and a cup of water.

At the start of his lesson, he asks his students to read the scenario in the probe activity. He then simulates an oil spill, pouring used cooking oil into the tray before asking his students to describe what they see, initiating the "ask" phase in the engineering design process. Several of his students make similar remarks, noting that the "dirty oil [contains] bits of food" and "[spreads] across the tray" when poured.

Teacher-directed scientific inquiry

Next, Mr Williams takes the mini mophead and uses it to clean up the oil spill. This time he asks, "How effective is the mophead in cleaning up the oil?" The students concur that the mophead cleaned up "some but not all of the oil". Mr Williams then dunks the mophead into the cup of water and attempts to clean the oil spill once more. Once again, he asks his students to describe what they see. One student remarks that the "oil spread further" when water was used, another shares that "smaller pools of oil" can be seen; a third student notices that "the oil and water did not mix", while a fourth student observes that "the mophead seems to absorb the water better than the oil".

Mr Williams is pleased with his students' descriptions and can sense their interest. Prior experience has taught him that his students respond favourably and remain engaged when a demonstration, role play, or game precedes an activity.

Keen to maintain his students' focus, Mr Williams keeps the pace by involving them in another activity. First, he lists the students' observations on the whiteboard. Then, he draws the bones of a mind map for oil and water and invites the students to fill in the properties of each liquid.

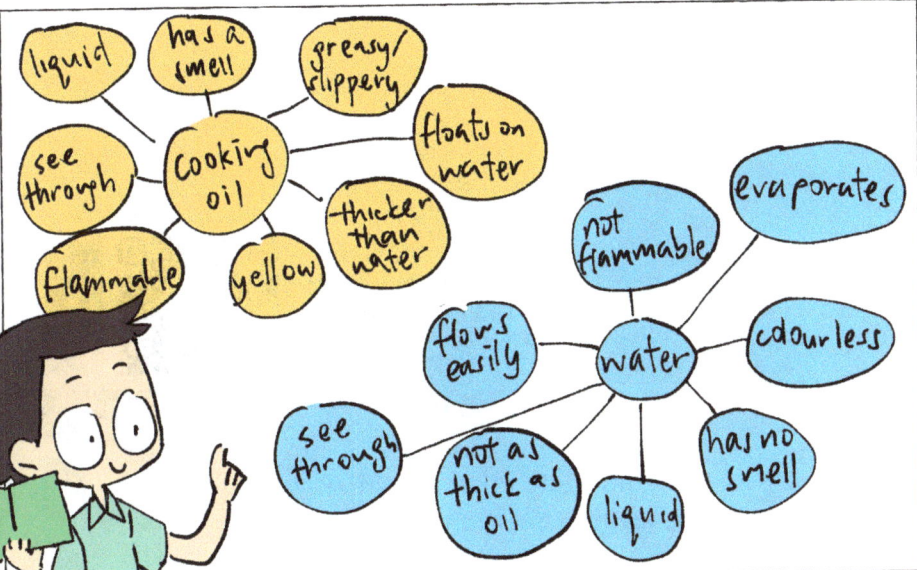

Teacher-directed scientific inquiry
Once the students have completed the mind maps, he diverts their attention back to their observations and asks, "Which of these properties help explain what you observed in the demonstration?" Here, Mr Williams uses a scientifically-oriented question to promote scientific discourse in the classroom. He first draws on the students' prior knowledge of mixtures, before exercising their deductive skills by guiding them to make connections between the contents of the mind maps with their observations.

Mr Williams draws the probe activity to a close by having his students form groups of four or five to engage them in a short brainstorming session to address the questions in the activity.

Mr Williams uses a demonstration to capture his students' attention and arouse their interests. Would this be effective in your classroom? What are some other ways you can spark your students' interests?

Scan this QR code to try out some topic-related riddles in your classroom!

Aside from mind maps, what other activities could you use to guide your students through the questions in the probe activity?

Many of the students reflected on their own experiences and observations from home. For example, one group compared different cleaning materials, proposing that, "tissue paper cannot soak up as much oil and water as cloth" and that "thicker material like paper towel would work better than the 'thread' used for the mophead". A few groups noted that Mateo, "should use detergent" or "soap", justifying that we "use detergent and soap to clean things in the kitchen".

In this vignette, Mr Williams uses the probe activity as a tool to scaffold his students' learning before introducing the real-life, persistent, and complex issue of oil spills in the environment. Mr Williams reveals the problem of interest in a series of steps, keeping his students on their toes. He first hands around a bag of puzzle pieces to each group. Each student picks four to five pieces and examines each piece carefully.

He then invites the students to exchange puzzle pieces with their group members and share their thoughts on what the final composite could be. In doing so, Mr Williams uses this activity to trigger the students' natural curiosity as they take turns to ask questions and offer guesses.

To keep his students intrigued, Mr Williams challenges them to a puzzle solving competition, drawing on their competitive spirits to keep them motivated and engaged.

This looks like it's part of a bird's body...

Look! When we put our pieces together, we can see that the bird is covered in oil!

Group problem solving

Once the puzzles are completed, Mr Williams invites his students to identify the common theme underlying each puzzle. To his surprise, they respond very quickly, some even before the puzzles are completed. His students later inform him that they were able to draw inferences from both the puzzle and probe activities. This was particularly encouraging for Mr Williams as he had intended to tap into various modalities of learning to draw on his students' critical thinking and logical reasoning skills.

Teacher-directed scientific inquiry

To develop these skills further, Mr Williams juxtaposes a picture of an oil spill at sea and an oil spill in the kitchen and writes key inquiry questions he would like them to consider on the whiteboard.

Some example questions include:

- In what ways are these two oil spills alike or different?
- Whom or what would be affected by the oil spills?
- Which oil spill would be easier to clean up and why?

Group problem solving

He gives the students 15 minutes to work in their groups to deliberate and record their answers in the form of a table.

Guidance to identify the problem

Making connections

During the group discussions, Mr Williams notes that a few groups could categorise their thoughts in a logical and orderly manner, using subcategories such as "similarities" and "differences", and "causes" and "effects". The remaining groups divided their answers into broader categories with labels akin to "oil spill at sea" and "oil spill in the kitchen". Mindful of the students' varied abilities, Mr Williams works with the answers from each group and uses scientific inquiry as cues to help his students categorise their answers more definitively. For example, he might ask, "What characteristic are you describing when you make observations on the look of an object?" or "What word describes how an object or substance feels like?"

Oil spill at sea	Oil spill in the kitchen
Dark Colour	Yellow Colour
Not see through	Clear (see through)
Dangerous to the animals and plants living beneath	Dangerous to people in the kitchen
Dangerous to people swimming near it	Easy to clean because of small volume of oil
Dangerous if it reaches beaches and land	Needs less manpower to clean up
Hard to clean up because of large volume of oil	
Needs more manpower to clean up	

Appearance — (circled: Dark Colour / Yellow Colour, Not see through / Clear)

Harmful to — (circled: Dangerous to people in the kitchen)

Harmful to — (circled left column: Dangerous to the animals..., Dangerous to people swimming near it, Dangerous if it reaches beaches and land)

As the students have prior knowledge on fuel types, their sources, uses, and the impact of Man on the environment, Mr Williams invites them to engage in a friendly class debate using the key inquiry questions and the answers they recorded.

Notably, Mr Williams challenged his students to consider:

- the implications that each scenario would have,

- the importance of understanding why and how to clean up each oil spill, and

- how the various STEM disciplines contribute to the process.

Take a moment to consider Mr Williams' key inquiry questions. How would you adapt these questions to accommodate students from a:

(a) Higher grade?

(b) Lower grade?

Alternatively, what other scientifically-oriented questions would you ask to encourage critical thinking and scientific discourse?

Part Two:
Research, Imagine & Plan

Next, Mr Williams presents his students with a starter kit consisting of plastic containers, equal volumes of water, equal volumes of two types of oil (one lighter and less viscous than the other, e.g., baby oil and vegetable oil), and various household items and materials including spoons, sponges, cotton balls and pads, detergent, and lemon juice.

After taking his students through the items in the starter kit, Mr Williams presents the students with the STEM challenge below.

Oil clean up starter kit

oil no. 1

oil no. 2

plastic spoons

water

cotton balls

sponge

Presenting a problem in a real-life context

Integrated STEM Activity

You are part of an oil spill research team and have been tasked to devise a solution to clean up two different oil spills at sea. Use the materials provided in your starter kit to design one or more prototypes to clean up the oil spills as <u>quickly</u> and <u>safely</u> as possible to prevent further spreading.

Guidance to identify the problem

Teacher-directed scientific inquiry

To encourage his students to use and develop their analytical skills, he invites them to deliberate on what they might need to know before diving head-first into the research process.

Examples of Mr Williams' guiding questions:

- What information would you need to find out how to clean up (1) an oil spill with a lighter and less viscous oil, and (2) an oil spill with a heavier and more viscous oil?

- Would you use the same or different types of materials to clean up each oil spill and why?

- Would you use the same methods to clean up each oil spill and why?

Student-centric inquiry

Group problem solving

The students then use Mr Williams' guiding questions along with others they have listed as the starting points of inquiry for their research, providing their research process with some structure. Mr Williams also reminds his students to explore resources beyond those available online, allowing students to visit the school library or bring along other resources where possible.

Before dismissing the class, he informs the students that they will be given some time to consolidate their research in the next lesson, before competing to build, test, and make improvements to their prototypes in the last lesson. To stimulate the students' competitive spirits and raise excitement, he informs them that the winning group will be awarded a prize, and that more points will be awarded to them.

What materials would you include in the starter kit for your students?

Classroom Tidbit
Rather than providing everyday materials in the starter kit, you may wish to challenge your students to explore materials that they are less familiar or unfamiliar with. You may also wish to encourage students to bring additional materials of their own.

In this STEM activity, Mr Williams challenges his students to clean up two types of oil, a lighter and less viscous oil, and a heavier and more viscous oil. Why do you think this is so? What types of oil(s) would you task your students to clean up and why?

Scan this QR code to access an editable list of materials for this activity.

Communication

Group problem solving

Engaging in iterative design

In the second lesson, Mr Williams gives the students 35 minutes to share their research findings and problem solve as a group.

During this stage, he guides students to better understand that the research, brainstorming and planning processes do not occur linearly, but rather, cyclically; he demonstrates this by inquiring about their ideas and plans, welcoming evidence-based justifications from the students, and referring them back to the "drawing board", to refine their inquiry questions and keyword searches further, so as to gain more in-depth knowledge on the specific materials or methods involved in an oil spill cleanup.

Communication

Prioritising evidence

Next, Mr Williams holds a quick sharing session, in which students from different groups talk about the different materials they have collected and justify why they are useful. He further engages the students by having them compare and explain why one material could be more effective than another.

paper towel

dishwashing pad

Making connections

Mr Williams also seizes the opportunity to explain how the knowledge and practices of the STEM disciplines can better inform solution design — that understanding the properties of materials (chemistry) can help us design more time and cost-efficient (mathematics) methods for oil cleanup (engineering and technology).

During the last 25 minutes of the lesson, Mr Williams gives each group three small containers: two containers with equal volumes of water and one of each oil type, labelled "Oil A" and "Oil B" respectively, and an empty container, labelled "Trash". He informs the class that the containers are to be used to test and evaluate their materials as they decide on a solution. Lastly, he reminds his students to record their observations as evidence to support their choice(s).

Rather than have his students test the materials at the prototype evaluation phase, Mr Williams gives his students some time to explore the materials during the imagine and planning phases. What are some advantages of this decision? How would you design these phases in your lesson plan?

Part Three:
Create, Test & Improve

Mr Williams divides the last lesson into two parts: **(1)** creating, testing, evaluating, and improving their prototype(s) (60 minutes), and **(2)** a final evaluated presentation (competition, 60 minutes).

Mr Williams plays into the students' competitive spirits by having them decide on the winning prize for the final presentation at the start of the lesson.

Teacher-directed scientific inquiry

He then observes the students as they go about constructing and testing their prototype(s). During this time, Mr Williams discovers that most of the groups have chosen to focus on the property of material absorbency. Rather than offer other suggestions for the students to consider, Mr Williams uses comparative questioning to guide students to make inferences on the role that other factors such as the prototype's shape and filtration ability may play.

For example: Which of the following shapes would help absorb more oil, a long and wide surface, or a short and narrow surface?

Peer evaluation

Mr Williams also involves the students in the assessment criteria decision-making process, to help his students develop a better appreciation of the evaluation process and take greater ownership over their prototype design(s). Much to his delight, his students request for some time to reevaluate their prototypes after deciding on the assessment criteria.

Marking Criteria

Ability to absorb a large amount of oil

Ability to remove the oil quickly

Ability to remove the oil without absorbing water

Ability to be used repeatedly

Mr William thus uses this opportune moment to reinforce that:

Prototyping is an iterative process. Innovation and progress often result from rigorous research, testing, evaluation, and redesign.

As large volumes of oil can be released into the waters during an oil spill, you may wish to invite your students to consider ways in which their prototype(s) can be used repeatedly to reduce material wastage. What are some other factors you would like your students to consider when creating their prototype(s)?

Classroom Tidbit
You may find that some students have trouble assessing the absorbency of each material. Depending on their grade level, they may not be familiar with many separation techniques. You may wish to demonstrate a simple technique such as filtering the oil and water mixture through a paper filter.

For the final part of the STEM activity and the last phase of the engineering design process, "communicate", Mr Williams requests for each group to present the key ideas behind their design choices using science-based evidence. This includes an explanation of why a group has chosen to make more than one prototype or use different materials to clean up the different oil spills. A representative is then selected to demonstrate the efficacy and effectiveness of their prototype(s) through a competition.

During the competition, each representative is given a mission to clean up equal volumes of light and heavy oil that have been "spilled" into equal volumes of water and then into two glass trays within 30 seconds. To assess the winner, the amount of oil and water removed is determined by filtering the mixture and calculating the percentage of oil/water removed over the total volume of oil/water used.

After each presentation, the remaining groups note down feedback onto a class feedback sheet for the presenting group to review and reflect on.

Rod can be turned to reel in the oil absorbent sheets

Mr Williams also takes some time to highlight: **(1)** the features that best fulfill the assessment criteria, **(2)** one or more aspects that could be improved, and **(3)** innovative ideas or prototype features.

Oil absorbent sheet

For example, while one group constructed a simple make-shift filtration system, another cut strips of oil absorbent material that could be collectively swept across an oil spill and reeled in. Notably, Mr Williams was impressed that his students did not introduce dispersants in their solutions. Instead, they took care in creating prototypes that could return the filtered water back to sea.

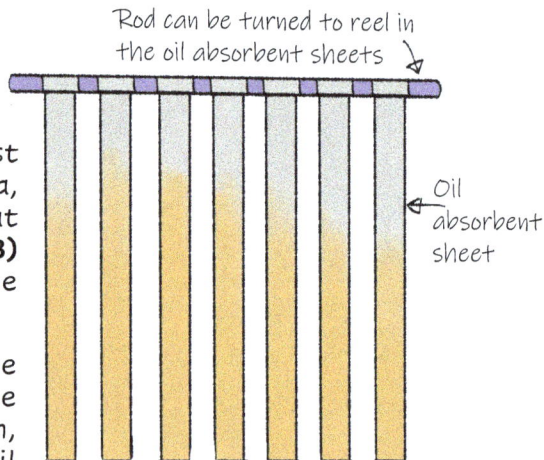

Mr Williams draws on his students' competitive nature by holding a competition. He uses the constraints of time and the ability of the prototype(s) to filter water out or return water back to "sea". What constraints or rules would you use?

What are some challenges you can foresee your students having during this integrated STEM activity? How can you help them overcome these challenges?

Scan this QR code to access the student workbook for this activity.

Mr Williams' Instructional Model

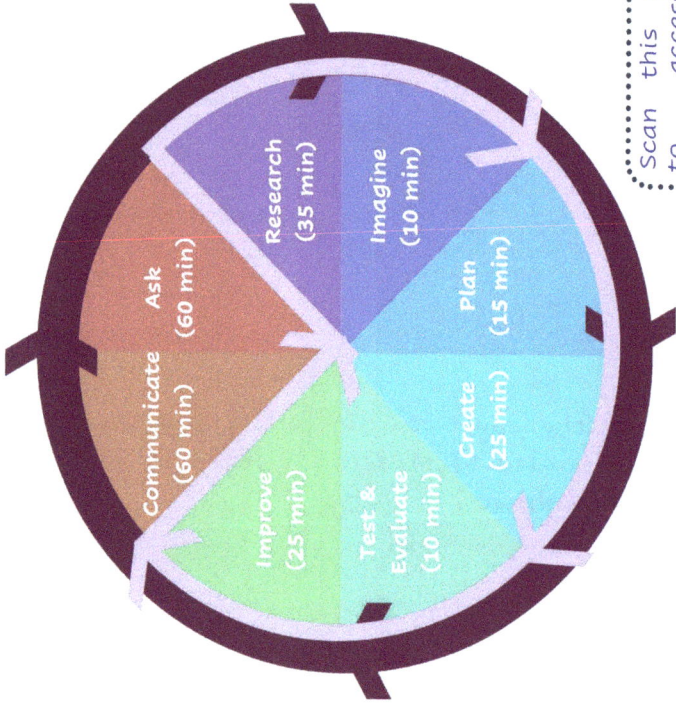

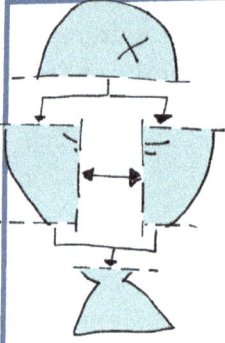

Research (35 min)

Imagine (10 min)

Ask (60 min)

Plan (15 min)

Communicate (60 min)

Create (25 min)

Improve (25 min)

Test & Evaluate (10 min)

Scan this QR code to access the detailed and editable instructional model.

Instructional Strategies

Students will:

Grade 11–12

Page 77
it is!

- Understand the chemical properties of different types of oils
- Recall different methods to clean up an oil spill
- Understand the underlying chemistry of different oil spill clean up techniques
- Design an experiment to simulate rough waters and heavy wind conductions at sea to test oils of different heaviness and viscosity
- Devise cleanup techniques to solve other consequences of an oil spill (e.g., dense tar mats on the seabed)

Classroom Tidbit
Unsure of how to adapt Mr William's oil spill activity for students of different grade levels? Pick your grade level, then follow the instructions.

Grade 9–10

Skip to
page 76!

- Understand mixtures and the properties of their constituents (e.g., immiscibility of oil and water)
- Learn about the effects of weathering on different types of oil (in particular, dispersion and emulsion)
- Explore different methods to clean up an oil spill and their drawbacks
- Design an environmentally less harmful dispersant/emulsifier to clean up an oil spill

Grade 7–8

Off to
page 75!

- Recall examples of different separation techniques
- Explore different methods for oil clean up
- Explore how two different types of oil respond to the weathering conditions at sea
- Explore ways to clean up an oil spill at sea under conditions of rough waters and heavy winds
- Design a cost-effective prototype to clean up an oil spill given the limitation of manpower

Grade 5–6

Back to
page 55!

- Understand the impact of fossil fuel consumption on the environment
- Understand how an oil spill can cause lasting adverse effects on the environment
- Learn about different types of oil and their uses
- Explore ways to clean up two different types of oil spills
- Design one or more prototypes to clean up each oil spill

Grade 3–4

Flip the
page!

- Understand the impact of water pollution on the Earth's environment
- Learn about other types of pollution and how they affect the environment
- Explore ways to clean up and prevent an oil spill from spreading further

In grades 3–4, students briefly learn about the impact of water pollution on the Earth's environment and that water is an important and limited resource. They can be guided to understand that an oil spill at sea can pollute the water and harm marine life. Students can similarly be introduced to other types of pollution for a more holistic learning experience. The probe activity can thus be used as an analogy for the importance of understanding how we can help protect our environment. The STEM activity in the featured vignette can be simplified into the following task.

Task: A truck delivering crude oil to a petrol station has crashed into a road barrier. Oil is seen leaking into the town's lake. As the town's fishermen depend on the lake for income, what can you do to save the lake from the oil spill?

Students can first learn about the different methods used to clean up an oil spill before using various everyday materials to simulate an oil spill cleanup of their own. At this stage students need not learn about the chemical composition of different oil types. Instead, they can be guided to appreciate the properties of different materials and how they interact with oil.

What would you adopt or adapt from this activity to suit the learner profiles in your classroom?

For the grade 7-8 classroom

At this level, students have learned about some separation techniques and can appreciate the importance of knowledge on the properties of substances. The probe can thus be used to spark a conversation about the properties of different oil types. Students can compare two or more types of oil (e.g., medium and thick crude oil and diesel) and learn how they respond under different weathering conditions at sea. Students can be guided to appreciate how knowledge of an oil's properties and its behaviour at sea plays a crucial role in selecting the cleanup technique(s) used. The STEM activity in the featured vignette can be modified into the following task.

Task: You are part of an oil spill response team tasked to clean up an oil spill at sea that is surrounded by rough waters and heavy winds. How can your team clean up the oil spill in these dangerous conditions?

Students can be asked to devise a prototype and technique that can be applied with the constraints of a limited budget and manpower. Here, students can be encouraged to devise semi-automated solutions made from recyclable materials.

What would you adopt or adapt from this activity to suit the learner profiles in your classroom?

At this level, students build further on their knowledge on how materials and substances interact with each other. The probe activity can be used to gauge students' prior knowledge on mixtures, particularly why oil and water are immiscible. Students can then be introduced to the effects of weathering on different types of oil and the drawbacks of some oil cleaning techniques before taking on a modified version of the grade 7–8 integrated STEM activity.

Task: It is the year 1989 and the world's largest oil spill has occurred at a site that is surrounded by rough waters and heavy winds. The response teams before you have failed to clean up the oil spill using traditional methods and time is of the essence. What other solutions can you devise to clean up the oil spill?

Students can be asked to: (1) find out how dispersion and emulsion differ, (2) conduct an experiment on how dispersants and emulsifiers can be used to clean up an oil spill, and (3) design an environmentally less harmful dispersant/emulsifier of their own. Students can then be further challenged to simulate the weathering conditions experienced at sea to test the effectiveness of their dispersant / emulsifier.

What would you adopt or adapt from this activity to suit the learner profiles in your classroom?

At this stage, students can appreciate the chemical properties of different oil types. The probe activity can hence be expanded upon to initiate a discussion on how different techniques and materials are used to clean up various types of oils. The integrated STEM activity suggested for a 9—10 class can then be modified into a case study to identify why two different oil spills can have very different environmental outcomes.

Case study: In 1999 an oil spill in the West of France resulted in a large-scale cleanup operation. In comparison, in 1993, twice the volume of oil was spilt at sea in Shetland, Scotland. Given that the sea conditions were similar, with rough and heavy winds, why did the 1993 oil spill produce a smaller economic and environmental impact compared to the 1999 oil spill? Design an experiment simulating similar conditions at sea to test oils of different heaviness and viscosity to explain what happened.

Dense tar mats can form from the combination of intense wave action and sediment stirred up into a water column. These tar mats can sink below the water surface or be swept onto shores. Students can be challenged to explore and devise cleanup techniques to remove the tar mats. Alternatively, students can explore several other consequences of an oil spill, such as formation of oil slicks, and devise a solution to solve it.

What would you adopt or adapt from this activity to suit the learner profiles in your classroom?

Reflection Questions

Now that you are half-way through your STEM journey, which aspects of STEM inquiry have you found most challenging to implement in your classroom?

What are some ideas, techniques or exercises you might like to try to better implement STEM inquiry when you conduct this activity?

How confident are you to carry out this activity in your classroom? Colour the fish to represent your confidence level and note down some reasons why.

Scan this QR code to access Mr Williams' editable lesson plan.

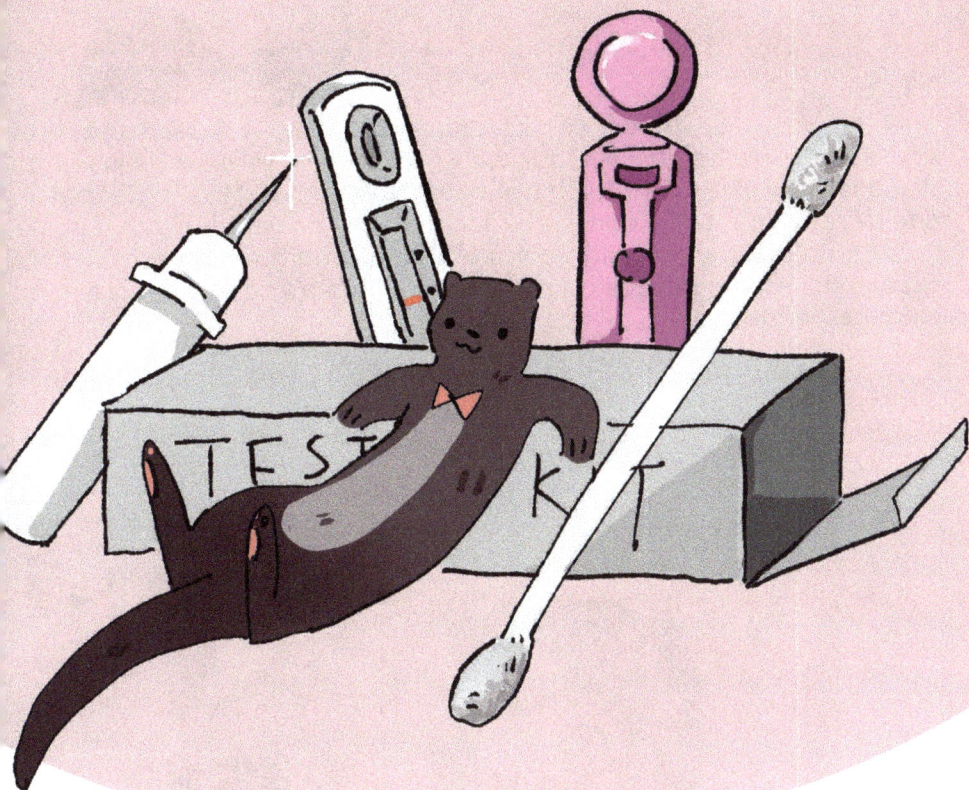

ACTIVITY FOUR
TEST KITS

What are test kits?

Probe Activity
Why Do We Need Test Kits?

This probe is designed to introduce commonly used test kits and screening devices in our daily lives. It can also be used to discuss how the intradisciplinary nature of STEM is reflected in the developmental process of such devices, particularly in a post-COVID, volatile, uncertain, complex, and ambiguous world, where fast, mass-scale detection and frequent and/or chronic screening may be needed.

Scenario

Sonia has recently been diagnosed with diabetes. She starts each morning by taking her blood glucose reading before breakfast. This morning, Sonia decides to share and explain her morning routine with Zara, her curious granddaughter who has come to visit.

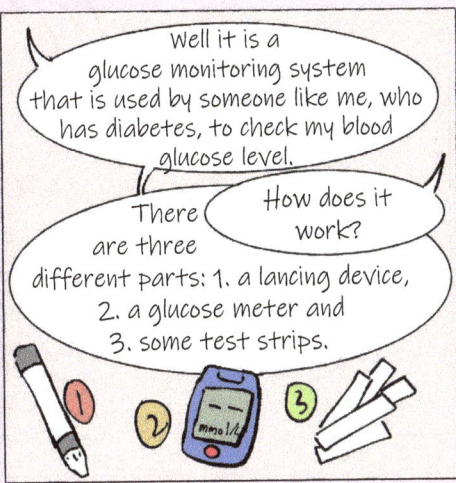

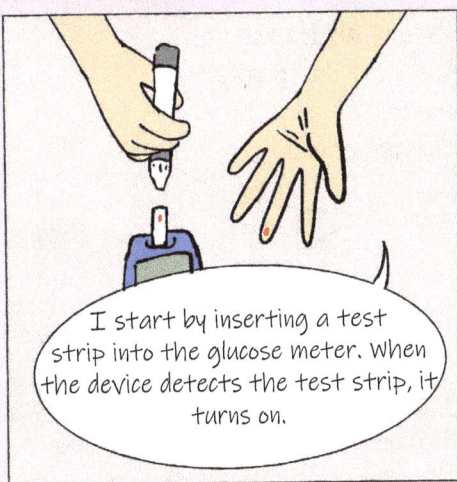

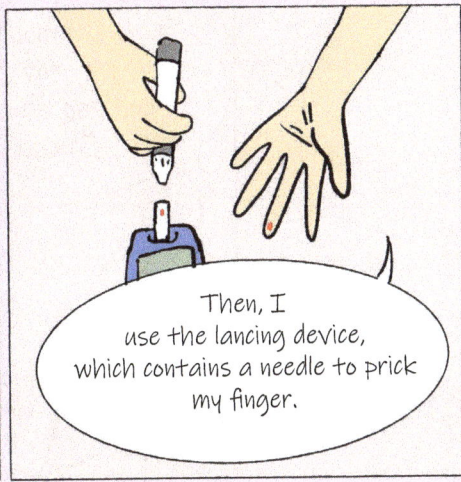

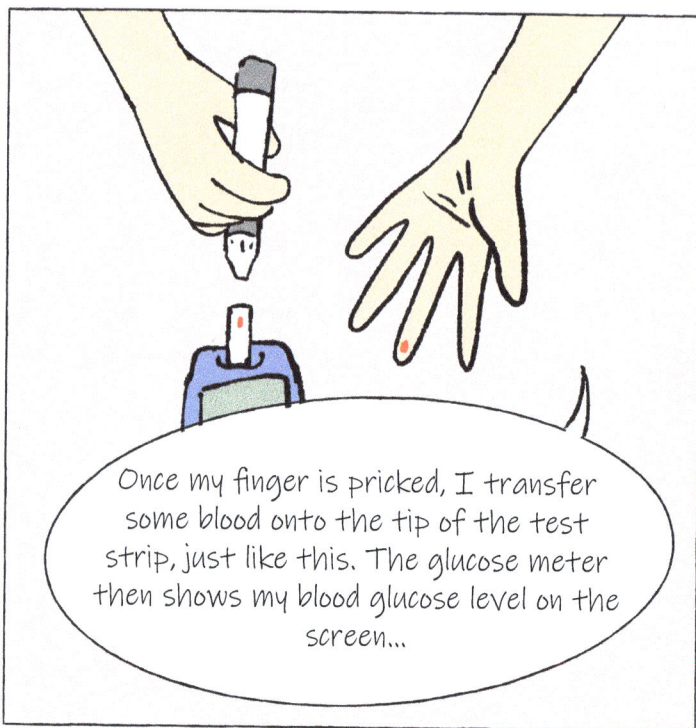

Have you seen or used a glucose monitoring system like Sonia's before?
This is one example of a test kit.
Why do you think a test kit like this would be useful in
(a) a clinic?

Classroom Tidbit
Try bringing a glucose monitoring system to the classroom for your students to observe more closely. Alternatively, show a video explaining the components of the glucose monitoring system and their functions.

(b) someone's home?

Classroom Tidbit
To answer this question, invite your students to consider the points in Zara's last question:
- Why would someone with diabetes need to know if their blood glucose level is normal?
- When would someone with diabetes need to test their blood glucose levels again?

What are some other examples of test kits?

What are some reasons why the test kits you have listed above are important?

The table below lists some relevant science concepts that can be covered in this integrated STEM activity on designing your own test kit.

Science Concepts/Grade Level	3—6	7—10	11—12
Adaptive and innate immunity			√
Biological molecules (carbohydrates)		√	√
Contributing factors in a pandemic			√
Human systems in the body and their functions	√	√	√
Infectious and non-infectious diseases		√	√
Nutrition and transportation in plants	√	√	√
Separation techniques (chromatography)		√	√
Structure of bacteria and viruses			√
The starch (iodine) test		√	√
Transmission and methods to reduce transmission of infectious diseases		√	√

What other science concepts or learning outcomes would you include in your lesson?

Implementing Integrated STEM Inquiry in the Grade 9—10 Classroom

The vignette on the next page offers insight into how "Why do we need test kits?" can initiate a conversation on: (1) the value of test kits, (2) the important features and functions of a test kit, and (3) important considerations when designing a test kit. The vignette describes how Mrs de Silva uses the PIRPOSAL instructional model and the flipped classroom approach to facilitate student-led, self-paced learning in her grade 9—10 classroom.

Integrated STEM Activity		
Activity	Task	Time (minutes)
Before Lesson One	Knowledge & Understanding	Self-paced learning
During Lesson One	Application & Analysis	90
Before Lesson Two	Design, Evaluate & Redesign	Self-paced learning
During Lesson Two	Create, Evaluate & Reflect	90

In the featured integrated STEM activity, students explore the concepts of infectious and non-infectious diseases outside of school hours and design and create a prototype of a test kit for a new emergent virus during school hours. Included in the description are examples of her students' responses and her reflections on the process. The characteristics of STEM inquiry relevant to her lesson are noted in the margin on the left.

Scan this QR code to watch a video on test kits.

Mrs de Silva has chosen to use the flipped classroom approach to deliver her lesson plan on infectious and non-infectious diseases. This approach centres around the idea that lecture-based learning and instruction is conducted before class, and class time is instead used to engage in application-based activities and higher-order thinking. As such, she has prepared all the necessary online resources for her students on Nearpod, an online teaching and sharing interface. Examples of the resources she has uploaded include: a PowerPoint video lesson on infectious and non-infectious diseases, a multiple-choice quiz on the video content, and additional resources to help scaffold, practise, and emphasise key concepts that her students will need to understand before undertaking the integrated STEM activity.

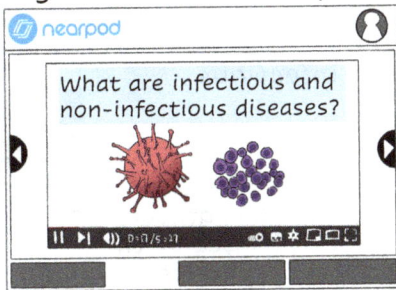

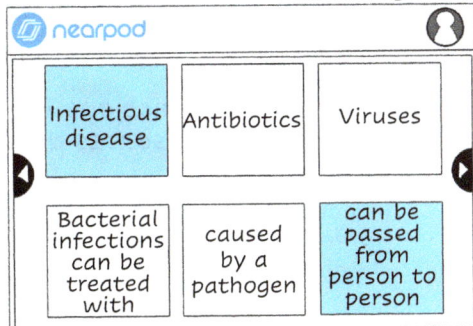

Having taught several of the students in her class before, Mrs de Silva is aware of the varied learning profiles in her classroom. She ensures that the additional resources are multimodal in nature, allowing students to apply the knowledge they have learnt in various ways, such as through fill-in-the-blanks and text-to-text matching activities.

With the intention to provide students with basic content knowledge using the learning materials, Mrs de Silva begins the integrated STEM activity with her class. She informs her students that the lesson materials can be viewed or attempted multiple times.

Student-centric inquiry

She takes the time to explain the concept of a flipped classroom and highlights the value of self-paced learning.

Furthermore, she encourages her students to explore the concepts, ideas, and information presented in the pre-lesson learning materials, on infectious and non-infectious diseases, more deeply in their own time.

Teacher-directed
scientific inquiry

Before the lesson begins, Mrs de Silva checks that her students have completed the pre-lesson materials using the "progress" function on Nearpod. She kickstarts the class by inviting her students to participate in a live discussion using Mentimeter, to gauge how much they have learnt.

During this discussion, students can post multiple, anonymous responses to the question, "How are infectious and non-infectious diseases different?" Mrs de Silva often uses live discussion tools to assess her students' prior knowledge on a topic or to gain instant feedback. As their responses appear in real time on the projector screen, Mrs de Silva is able to immediately identify any gaps in her students' knowledge and address them.

How are infectious and non-infectious diseases different?

Infectious diseases can spread, non-infectious diseases cannot.

One is caused by pathogens, the other is caused by the body e.g., genetics?

Bacteria and viruses cause infectious diseases

Non-infectious diseases are caused by things like environmental factors

Infectious diseases spread orally, through the air, touch, etc. but non-infectious diseases do not spread in these ways

15

Students are typing...

For example, when Mrs de Silva notices that the students were using "the flu" and "cold" interchangeably, she quickly debunks the misconception by explaining that the common cold and the flu are caused by different viruses. In addition, as this is a common misconception held by students, Mrs de Silva presents an online article from the Centre for Disease Control and Prevention for them to read. After giving her students some time, she asks them to summarise the article's main points on the whiteboard. In doing so, Mrs de Silva continues to promote student-centred learning in her classroom.

Mrs de Silva uses the flipped classroom approach to promote student-centred learning and agency in her classroom. Would this teaching style suit the learner profiles in your classroom, and why?

When students first learn about infectious diseases, they oftentimes misconstrue the influenza as the common cold. What are some other misconceptions you can anticipate your students having?

Classroom Tidbit
You may wish to highlight some of the misconceptions you have listed in more than one way, such as through pre-lesson notes or "Myths and Facts" and "True or False" activities, to enrich students' understanding and facilitate better knowledge acquisition of the topic.

For the most part, Mrs de Silva designed the pre-lesson materials and out-of-lesson tasks to encourage learner agency. For example, students are given a research activity prior to the first lesson, to learn about an infectious and non-infectious disease, and to share their knowledge in the form of a diagram on the "Infectious and Non-Infectious Diseases" Collaborate Board on Nearpod. Mrs de Silva imposes no specific criteria on the research task as she hopes that her students will practise self-led scientific inquiry and become active learners who harness an interest in exploring scientific content.

Student-centric inquiry

As Mrs de Silva also wishes to boost her students' confidence in self-agency, she invites them to work in pairs to share their findings from the research task. Students then swap partners and repeat the activity. In this way, Mrs de Silva uses this student-to-student teaching exercise to foster an environment in which students take ownership of their learning, shifting away from teacher-directed student engagement and towards student-centred, self-empowered learning.

Teacher-directed scientific inquiry

Next, Mrs de Silva divides the whiteboard into halves and holds a class discussion on what they have learnt from each other. She lists some of the infectious and non-infectious diseases named by her students on the left and right sides of the whiteboard respectively and makes mini mind maps using the information given. She then asks, "How are the diseases within each group similar?" This helps her students consolidate their knowledge by (1) applying the characteristics identified at the start of the lesson and (2) placing emphasis on the differences between infectious and non-infectious diseases.

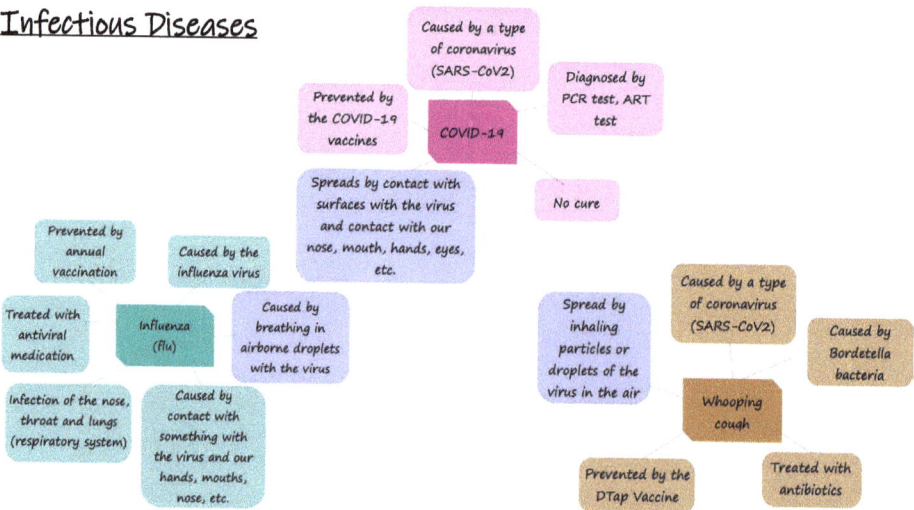

Infectious Diseases

COVID-19
- Caused by a type of coronavirus (SARS-CoV2)
- Diagnosed by PCR test, ART test
- Prevented by the COVID-19 vaccines
- Spreads by contact with surfaces with the virus and contact with our nose, mouth, hands, eyes, etc.
- No cure

Influenza (flu)
- Prevented by annual vaccination
- Caused by the influenza virus
- Treated with antiviral medication
- Caused by breathing in airborne droplets with the virus
- Infection of the nose, throat and lungs (respiratory system)
- Caused by contact with something with the virus and our hands, mouths, nose, etc.

Whooping cough
- Spread by inhaling particles or droplets of the virus in the air
- Caused by a type of coronavirus (SARS-CoV2)
- Caused by Bordetella bacteria
- Prevented by the DTap Vaccine
- Treated with antibiotics

90

Student-centric inquiry

In keeping with her "students as teachers" approach, Mrs de Silva brings the discussion to a close by having her students present and answer each other's queries, providing guidance and clarification only when needed. As most of her students have attempted the pre-lesson materials multiple times, Mrs de Silva can see that her students seem confident and are participating in the discussions more actively than usual.

Teacher-directed scientific inquiry

Given this, Mrs de Silva introduces the probe activity by first having her students read it. She then asks them to consider what test kits are and why they are used. Generally, her students express that test kits are used to test for "certain diseases or sicknesses" or "levels of things that can be harmful to us".

Group problem solving

Satisfied with their responses, Mrs de Silva divides the class into two teams. Each team is given 15 minutes to draft as many rebuttal points as possible on the topic of why test kits are more important for infectious diseases than non-infectious diseases.

Mrs de Silva then holds a 10-minute debate, in which each team can freely refute the opposition's points by using scientific evidence and/or scientific reasoning skills.

Prioritising evidence

Making connections

As a surprise, Mrs de Silva dons a black robe and arms herself with a toy gavel to chair the debate, infusing fun into the court-like enactment, to keep student engagement levels up and heighten their interests.

She uses this back-and-forth debate to:
1. help students understand the value and need for test kits in the diagnosis and management of some diseases,
2. guide students to answer the probe activity's questions, and
3. fortify connections between STEM knowledge and competencies and the development of everyday (medical) devices that are beneficial or useful to our health.

Mrs de Silva designed checklists to provide her students with structure and guidance to work systematically, outside of the classroom. What are some other ways you could help your students work systematically and more independently?

Teacher-directed scientific inquiry

Next, Mrs de Silva asks her students to recall how the starch (iodine) test and Benedict's test work. She then asks, "If the materials required to conduct the starch (iodine) test or Benedict's test have been prepared for you, would you consider these materials to be part of a test kit? How alike or different would these test kits be compared to the ones you have listed in the probe activity?" This not only helps Mrs de Silva reinforce that test kits can differ in both function and appearance, but also creates an opportunity for her to introduce her students to the concept of qualitative and quantitative testing and when each type of test might be used.

To bring the discussion to a close and assess her students' comprehension on qualitative and quantitative testing, Mrs de Silva selects a few students to name a test kit and justify if its function is an example of qualitative or quantitative testing.

Towards the end of the lesson, Mrs de Silva divides her students into groups of four or five before presenting the integrated STEM activity below.

Integrated STEM Activity

An alarming number of cases of a new influenza virus has been reported across the country. Considerable burden is being placed on the healthcare system to screen and diagnose patients. Your research group at the Centre for Disease Prevention and Control has been tasked to devise a test kit that can be self-administered for early detection and treatment.

The test-kit must be:
(a) light weight and portable,
(b) low in cost to produce,
(c) easy to use by a layperson, and
(d) accompanied by a brochure with instructions on how to use it.

Students are given the rest of the lesson to analyse the problem, brainstorm ideas, and search for information to better inform their solution choices.

While listening in on the group discussions however, Mrs de Silva notices two groups experiencing some trouble understanding content on adaptive and innate immunity and how the various COVID-19 test kits work. Aware that the students are learning this knowledge for the first time, she assists them by breaking down the unfamiliar content and use of scientific language into simpler terms and diagrams, sharing instructional videos with the class when needed.

Scan this QR code to access a list of materials that could be used to make a test kit prototype.

In this vignette, Mrs de Silva tries to enrich her students' learning experience and sense of self-empowerment through student-to-student teaching.

(a) Would this "students as teachers" strategy work in your classroom, and why?

(b) What strategies would you implement to cultivate self-empowerment and connectedness in your students' learning experiences?

Mrs de Silva believes that fun can help bolster student engagement and stimulate creativity and innovation. What are some ways you can balance "loosening up" your classroom with meaningful and deep learning?

Before Lesson Two:
Design, Evaluate & Redesign

Mrs de Silva uses Nearpod once again as the primary interface to upload her pre-lesson materials and communicate with her students outside of the classroom. Her students are given three checklists to guide them through the design, evaluation, and redesign processes before the second lesson: **(1)** a task checklist, **(2)** a checklist to facilitate student assessment of how the disciplines of STEM have been incorporated into their solution(s), and **(3)** a checklist with potential factors to consider when evaluating the design, cost, and user-friendliness of their solution(s).

Group problem solving

Prioritising evidence

Of the three checklists, (2) and (3) have been intentionally left incomplete. For checklist (2), students are required to work as a group to critically analyse and identify if the knowledge, competencies, and skills of two or more STEM disciplines have been incorporated into their solution(s). For checklist (3), students are required to identify key criteria for the design and function of a test kit and justify whether their solution(s) meet(s) the criteria. Mrs de Silva also encourages her students to explore unconventional ideas through suggestions in checklist (1). For example, when evaluating the different solutions, she invites students to consider blending components from two or more solutions.

Peer evaluation

Engaging in iterative design

The three checklists were hence designed to cultivate her students' skills in time management, collaboration, adaptability, flexibility, analytical thinking, and logical reasoning.

Lastly, to acquaint her students with the iterative cycle of prototype design, each group must conduct their own internal review and redesign processes prior to and after evaluating the designs of the other groups on Nearpod.

Scan this QR code to access the editable checklists!

95

Dropper

Extraction solution

Severity of the viral infection

Nasal swab

Sample application

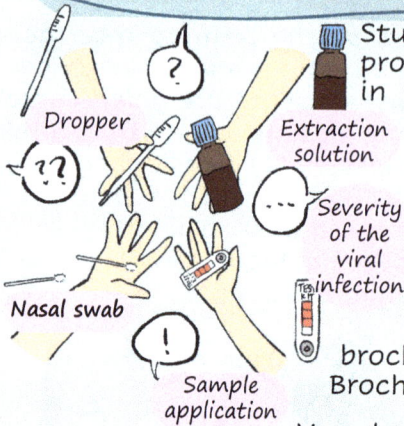

Students construct their test kit prototypes and work on their brochures in this lesson. Rather than taking on a supervisory role, Mrs de Silva observes her students from afar, offering guidance only when requested, allowing students to take ownership of the prototyping process. After completing their prototypes, each group uploads a video of their prototype and brochure onto the "Final Prototype and Brochure" Collaborate Board on Nearpod.

Mrs de Silva then presents them with an evaluation form. Having designed their own criteria checklist in "Before Lesson Two", a few students remarked that the evaluation form did not allow for other criteria to be added and requested to have the option of adding their own criteria. Pleased to see positive reactions from the other students, Mrs de Silva quickly amends the evaluation form.

Peer evaluation

Teacher evaluation

Students are then given 30 minutes to evaluate the prototypes and brochures of other groups before reflecting on how they can improve their own prototype. To guide her students in this final evaluation process, Mrs de Silva shares her criteria with her students, explaining the reasoning behind her choices. All evaluation and reflection forms are then uploaded onto Nearpod for the students to review.

To bring the integrated STEM activity to a close, Mrs de Silva holds a live anonymous discussion on Mentimeter. Students can cast individual votes for the best prototype and brochure and openly express why. She then hands out certificates of completion. The certificates are personalised and feature the skills or exercises in which a student excelled in. She plans to give out certificates for each integrated STEM activity and hopes to hold a science fair to showcase her students' work. Students will work towards a final award of completion, which will be presented at the science fair. The lesson ends with feedback from her students on their flipped classroom experience.

Scan this QR code to access the evaluation form.

Students are often evaluated by their teacher and/or peers.

(a) What are some other ways you can provide useful feedback to help your students become better designers and problem solvers?

(b) Feedback is typically given in a verbal or written form. What are some other non-verbal ways your students can receive feedback to make the evaluation process more interesting and interactive?

Scan this QR code to access the student workbook for this activity.

Mrs de Silva's Instructional Model

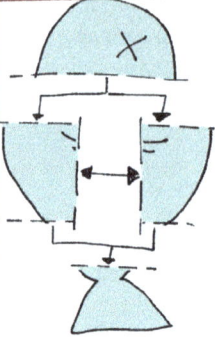

Teacher's Tasks

Students' Tasks

Before Lesson One: Knowledge & Understanding

During Lesson One: Application & Analysis

Before Lesson Two: Design, Evaluate & Redesign

During Lesson Two: Create, Evaluate & Reflect

Scan this QR code to access the detailed and editable instructional model.

Instructional Strategies

Students will:

Grade 11–12

Page 103 it is!

- Describe the components of various test kits and their functions
- Compare and deduce what qualitative and quantitative tests are and give some examples
- Create a prototype that simulates how a test kit for a newly emergent virus might work
- Suggest how to package and store a test kit and its components

Grade 9–10

Back to page 86!

- Describe how various test kits are used
- Learn about qualitative and quantitative testing
- Identify common and unique components of qualitative and quantitative test kits
- Design a prototype test kit and brochure for a newly emergent virus

Grade 7–8

Off to page 102!

- List everyday examples of test kits and describe their functions
- Describe key components of some test kits
- Learn about the starch test kit and how to use it
- Design an experiment to test for the presence of starch before and after cooking different types of food

Grade 5–6

Skip to page 101!

- Learn about everyday examples of test kits and why we use them
- Learn about key components of some test kits
- Learn about the starch test kit and how to use it
- Design an experiment to test the presence of starch in different types of food

Grade 3–4

Flip the page!

- Learn about everyday examples of test kits and why we use them
- Learn about the starch test kit
- Experiment testing various foods and materials using the starch test kit

> **Classroom Tidbit**
> Unsure of how to adapt Mrs de Silva's test kit activity for students of different grade levels? Pick your grade level, then follow the instructions.

At this level, students may find terms like bacteria and viruses difficult to understand. You may wish to conduct simple hands-on activities for students to explore how "germs" or "microorganisms" spread and how to prevent them. The probe activity can be used to introduce non-infectious diseases. Students can be guided to understand basic differences between infectious and non-infectious diseases. Due to the COVID-19 pandemic, students may know of rapid antigen test kits. You may wish to show how such a test kit works in comparison to a glucose monitoring system, to discuss different types of test kits and their uses. Students can then be given the integrated STEM activity below.

Task: Abigail has been diagnosed with diabetes and was advised to reduce her intake of starchy foods. However, she is unsure which types of food contain starch. She wishes to find a way to test her meals. Use the starch (iodine) test to find out, "How can you help Abigail test and avoid the starchy foods in her meals?"

Here, students can learn about the starch (iodine) test and how to use it. You may wish to select foods that either contain or do not contain starch so that the differences can be clearly observed. Students can then find out about foods that contain starch and assess whether their daily meals are suitable for Abigail. This helps raise awareness of nutritional diseases and emphasises the value of home-based test kits.

What would you adopt or adapt from this activity to suit the learner profiles in your classroom?

For the grade 5-6 classroom

In grades 5-6, students have learned about the circulatory system. The probe activity can be used to discuss the causes of diabetes and how it can affect the circulatory system. Students can then be asked to compare diabetes to the common cold/COVID-19 (and how it is spread) to spark a conversation on the differences between infectious and non-infectious diseases, and how test kits are useful. They can be guided to learn about the various components and functions of selected test kits before learning about the starch (iodine) test kit. Students can then be tasked to carry out a modified version of the integrated STEM activity for grade 3-4 students.

Task: Abigail has been diagnosed with diabetes and was advised to reduce her intake of starchy foods. However, she is unsure which types of food contain starch and wishes to find a way to test the groceries she bought. Use the starch (iodine) test to design an experiment to find out, "How can you help Abigail test and avoid starchy foods?"

Students can be challenged to deduce why some foods have a darker bluish-purple stain than others despite using the same amount of iodine solution. This can help them gain an appreciation of how a basic test kit works and introduce them to the concept of how the starch (iodine) test can be both qualitative and quantitative in nature.

What would you adopt or adapt from this activity to suit the learner profiles in your classroom?

At this level, students learn about the digestive system, transport systems and diffusion. The probe activity can be used to introduce the importance of glucose in the human body and how it is derived from our diet. Students can find out how glucose is transported to the cells in our body (passive and active transport) and discuss the importance of monitoring our blood glucose level. The second question in the probe activity can then be used to introduce students to the starch (iodine) test kit and initiate a discussion on the similarities and differences between various test kits. This can help students appreciate how test kits vary in complexity depending on their function. Students can then be given a modified version of the STEM activity for grade 5–6 students.

Task: Abigail has been diagnosed with diabetes and was advised to reduce her intake of starchy foods. To maintain a healthier diet, she has decided to eat more vegetables. However, she is unsure whether raw or cooked vegetables would be more beneficial. Does the amount of starch change before and after cooking? Use the starch (iodine) test to find out, "How can you help Abigail decide if raw or cooked vegetables are better for her?"

Students can be challenged to explore the starch content of other foods that can be eaten raw. They can be further challenged to learn about resistant starch and its benefits.

What would you adopt or adapt from this activity to suit the learner profiles in your classroom?

For the grade 11-12 classroom

By grades 11–12, students have learned about infectious and non-infectious diseases. As such, the probe activity can be used to have students recall prior knowledge on various test kits and their functions. Students can be tasked to compare different test kits and their components and be guided to identify that some test kits are qualitative, while others are quantitative in nature. This can prompt a discussion on where qualitative and quantitative tests might be useful. The featured STEM activity in the vignette can then be modified as follows.

Task: An alarming number of cases of a new influenza virus has been reported across the country. Considerable burden is being placed on the healthcare system to screen and diagnose patients. Your research group at the Centre for Disease Prevention and Control has been tasked to devise a test kit that can be self-administered for early detection and treatment. The test kit must be: (1) lightweight and portable, (2) low in cost to produce, (3) easy to use by a layperson, (4) accompanied by a brochure with instructions on how to use it, and (5) able to simulate a positive and negative result.

Students can be further challenged to consider how they can package and store their test kits: (1) using environmentally friendlier options, and (2) to ensure longer shelf-life.

What would you adopt or adapt from this activity to suit the learner profiles in your classroom?

Reflection Questions

Take a moment to reflect on the flow of STEM inquiry in the activities you have read. Have you noticed that the features of STEM inquiry do not always occur in a linear fashion?

(a) Describe your STEM inquiry experience in your classroom.

(b) Do the features of STEM inquiry have to flow in a linear fashion? Express your thoughts and reasoning(s) in the space provided below.

How confident are you to carry out this activity in your classroom? Colour the fish to represent your confidence level and note down some reasons why.

Scan this QR code to access Mrs de Silva's editable lesson plan.

ACTIVITY FIVE
THE NEUROMUSCULAR SYSTEM

What role does the neuromuscular system play in the human body?

Probe Activity
How Do Our Muscles Contract?

The probe activity below can be used to introduce the concept of neuronal signalling and activation of muscle fibres in the neuromuscular system. It is designed to help students understand how systems can interact, and that the systems in our body work together to carry out tasks, such as coordinating movement.

Scenario

Priya's older brother, Ankush, has returned from his evening jog. He fetches a device from the cabinet and sits on the living room sofa. He attaches two sticky pads on either side of his knee before turning the device on. Having seen this device for the first time, Priya could not help but be curious.

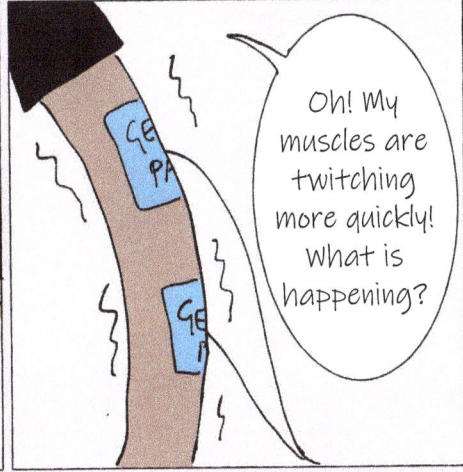

Perhaps you have seen a device like this before. What do you think it is used for?

Suggest which system(s) in your body the device and scenario are mimicking and why.

The table below lists some relevant science concepts that can be covered in this integrated STEM activity on the neuromuscular system and neuromuscular disease.

Science Concepts/Grade Level	3—6	7—10	11—12
Crossbridge cycling (sliding filament theory)			√
Interacting systems in the human body	√	√	√
Muscle relaxation and contraction		√	√
Muscle types and functions	√	√	√
Neuromuscular disorders and dysfunction	√	√	√
Neuroplasticity			√
Neuromuscular system		√	√
Neuronal communication (signalling)			√
Nervous system (central and peripheral)	√	√	√
Voluntary and involuntary actions (reflex arc)		√	√

What other science concepts or learning outcomes would you include in your lesson?

Implementing Integrated STEM Inquiry in the Grade 3—4 Classroom

The vignette below offers insight into how "How do our muscles contract?" can facilitate STEM inquiry on the neuromuscular system and neuromuscular impairment in a grade 3–4 classroom.

The vignette features an integrated STEM activity conducted by Mr Chen as a workshop. He adopts the Blended Socratic Method of Teaching to integrate biology as the lead science discipline and engineering to encourage his students to design solutions that can improve mobility in a stroke rehabilitation patient.

Integrated STEM Activity		
Part	Task	Time (hours)
One	Introduction	
Two	Research, Design & Prototype Proposal	
Three	Creation & Testing of the Prototype	6.0
Four	Prototype Presentation & Evaluation	

Included in the description are examples of his students' responses and his reflections on the process. The characteristics of STEM inquiry relevant to his lesson are noted in the margin on the left.

Scan this QR code to watch a video on how the neuromuscular system can be introduced.

Part One: Introduction

As the students enter the workshop, they find Mr Chen seated at the teacher's desk with a device in one hand and two square pads, each connected to a lead, stuck onto his other forearm. Intrigued by the scenario, his students are quick to question, "Mr Chen, what are you doing?", "Mr Chen are you okay?", "Mr Chen what is that?"

Taking their curiosity as his cue to proceed, Mr Chen enacts his rendition of the probe activity before asking, "Here, why don't one or two of you try this out!"

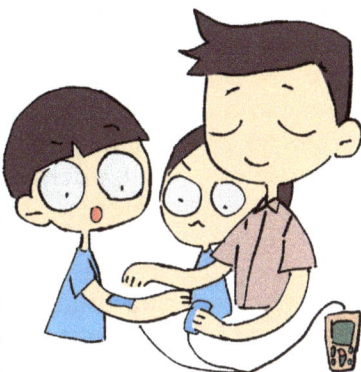

After placing the sticky pads onto the forearm of a volunteer, he adjusts the setting to a medium intensity and asks, "Can you describe how you feel?" He then repeats the demonstration on another volunteer. While the first student exclaimed that they could feel their "muscles twitching", the second student notes that they could feel "funny buzzing tingles". To Mr Chen's surprise, one of the observing students shares that they could "see the skin around the sticky pads move".

Noting their observations on the whiteboard, Mr Chen asks, "Have any of you seen a device like this before? From your classmates' descriptions, what do you think this device is used for?" A mixture of responses could be heard. While some are quick to reply that the device is "for massaging sore muscles", others respond that the device "is used in the hospital".

At this grade level, though his students have some knowledge of the systems in the body, they have yet to learn of the neuromuscular system. As such, he uses strategic, scientifically-oriented questions about movement to scaffold their learning and introduce the neuromuscular system.

Teacher-directed scientific inquiry

He begins by acting out a series of movements, using questions to help his students establish a connection between the control centre of the body (the brain) and the parts necessary to enact a particular movement.

For example:

- What parts of my body am I using to swing my arm from left to right?
- What do we have in our arms and legs that help us move?

As his students have recently learned about the digestive system, Mr Chen uses their prior knowledge to guide them to apply the concept of systems level pathways to the movements they have observed. He begins by noting down the path of food through the digestive system. As expected, his students do not hesitate to chime in and complete the pathway on the whiteboard.

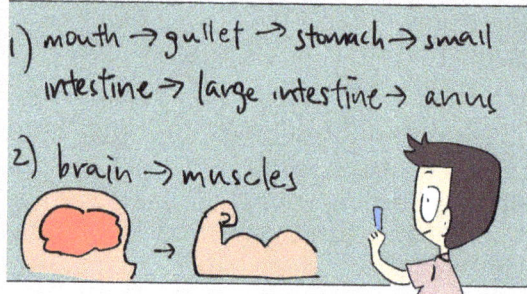

1) mouth → gullet → stomach → small intestine → large intestine → anus

2) brain → muscles

Teacher-directed scientific inquiry

Mr Chen then draws their attention to a second pathway, "brain -> muscles in arms/legs". Pointing to the whiteboard, he challenges his students to consider:

If the digestive system describes how food travels through the body, then what do you think a system that helps us move is called?

When his students cannot identify the name of the system, Mr Chen takes this opportunity to introduce the neuromuscular system as a network of nerve cells (neurons) that extend from the brain to the muscles in our body, sending signals that tell the muscles to contract.

Making connections

He then adapts the second probe activity question by asking, "Given what you now know, which parts of the neuromuscular system do you think matches this device and what you observed in my demonstration earlier?"

To help his students along, Mr Chen draws a simple schematic on the whiteboard and asks students to match the words "brain", "nerves" and "muscles" to the diagram. In doing so, Mr Chen deliberately uses a visual analogy to help his students understand how the neuromuscular system works on a broad level.

Students in higher grades may be familiar with electrical muscle stimulation devices.

(a) Would a demonstration like Mr Chen's be well-received in your classroom? How would you introduce the neuromuscular system differently?

(b) In what ways can you expand on the questions in the probe activity to adapt to the content knowledge of higher grade level students?

Classroom Tidbit
Grade 9–12 students are familiar with the concept of coordinated responses. You may wish to adapt the questions in the probe activity to gauge their prior knowledge on voluntary and involuntary responses.

Next, Mr Chen shows a video of a stroke patient being assisted in moving from a hospital bed to a wheelchair. He guides his students to understand the context of the video through observation-based inquiry.

For example:

- Can you describe the people in this video?

- Where do you think the people are and why?

- What is happening in this video?

- What do you think has happened to this patient?

Upon hearing his students' observations that the patient "cannot move", and "might have been in an accident", he asks, "Who has heard of a stroke before? Can you tell me what it is?" Aware that his students are in grade 4 and may not have the content knowledge to answer the question, Mr Chen breaks the broad question down into smaller, more manageable questions.

For example:

- Remember the pathway, "brain -> muscles in arms/legs"? Which body part or organ controls movement?

- What would happen if the part of your brain that is responsible for controlling the movement of your fingers was damaged?

In doing so, Mr Chen guides his students to understand some of the signs and consequences of a stroke without simply offering an explanation. Confident that his students understand the context, he presents them with a challenge.

Integrated STEM Activity

John has been hospitalised with a mild stroke. He has lost muscle strength in the right side of his body and is experiencing muscle stiffness, particularly in his right arm. As a rehabilitation engineer, you have been tasked to assist John with his recovery. How can you design a device that can help John regain movement in his right arm?

To help his students better understand the context of the problem, Mr Chen gives each student access to a PowerPoint on the neuromuscular system and neuromuscular disease, and a video on the consequences of stroke to help further reinforce the points discussed in the lesson. Students are then given approximately half an hour to work through the slides and video individually on their iPads.

Concerned that the topic may be a little difficult for his students, Mr Chen engages them in a quick Kahoot quiz, to assess the depth of their newly acquired knowledge and determine if any areas require further explanation.

19 Kahoot! **3** Answers

Numbness in the arm | Problems in keeping your balance
Trouble speaking | All of the above

He ensures to also give students the opportunity to express themselves openly and anonymously through an online platform, encouraging them to share their questions and challenges through anonymous posts.

Scan this QR code to access PowerPoint slides on the neuromuscular system and neuromuscular disease.

115

Mr. Chen uses technology to conduct his workshop in several ways: (1) as a self-paced learning tool, (2) as an assessment tool, and (3) as a communication tool. How would you implement the "T" in "STEM" in your classroom?

Part Two: Research, Design & Prototype Proposal

Guidance to identify the problem

Following the anonymous question and sharing session, Mr Chen brings the students' focus back to the challenge. He randomly divides the students into groups of five by having them draw numbers from a bag.

He then invites them to work in groups to identify:

1. What has happened to John?

2. What issue is John facing?

3. What are they tasked to do to help John?

After informing them that they will be given two hours to research, design, and present a solution in the form of a poster presentation, the students are given 15 minutes to plan how they will use their time. Each group also nominates a timekeeper to ensure that they keep to the two-hour limit.

Mr Chen also unveils shelves of materials at the back of the classroom, inviting students to freely explore, be inspired by and use any of the materials they find over the next two hours.

Group problem solving

As the students research how to increase muscle strength and improve arm mobility, Mr Chen listens to and observes them carefully. As he wishes to encourage his students to ask scientifically-oriented questions, he reminds them of the anonymous question platform and keeps it open throughout the workshop. Occasionally, he calls a timeout to address their queries.

During his observations, Mr Chen notices that different groups are focusing on different aspects of the integrated STEM challenge. While most groups are focused on strength training, two groups are focused on creating devices that can help a patient lift or fetch an object more easily. As the time passes, Mr Chen also notices that the groups designing strength training devices have mostly designed makeshift weights and are well ahead of the other two groups.

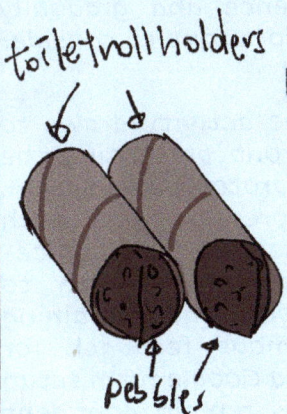

To encourage these students to **think more creatively** and **innovatively**, Mr Chen challenges them to devise an **alternative**, **unconventional** way for a stroke rehabilitation patient to practice lifting weights:

> How can you modify your solution to make the rehabilitation activity more interesting or fun for the patient?

117

Engaging in iterative design

Group problem solving

To his surprise, one group is quick to propose a fishing rod-inspired solution, whereby a weight can be attached to the end of a rod and lifted as the patient rotates the rod with their hands.

Encouraged by their idea, Mr Chen challenges them to take the design process one step further. He invites them to consider ways to modify their design to support heavier loads without causing the patient too much strain.

At this stage, although students have not begun to construct their prototypes, Mr Chen reminds them that they can source and test out the various materials provided at the back of the classroom. He does this deliberately to cultivate an environment of exploration and instill a questioning mindset.

Teacher evaluation

This is an example of how Mr Chen uses the feedback process to scaffold his students' STEM learning experience and gradually bring them out of their comfort zone to consider more complex and user-centric ideas.

Communication

Peer evaluation

Prioritising evidence

The second part of the activity draws to a close with each group presenting the proposal for their prototype's design. During their poster presentations, each group must provide science or evidence-based observations and/or research to support their decisions. The remaining groups leave anonymous feedback for their classmates using a Google form setup by Mr Chen. To help his students remain confident and open to the design evaluation and revision process, Mr Chen takes a moment to highlight examples of positive and constructive comments. The students are then given 30 minutes to discuss and revise their designs based on the feedback they have chosen to focus on from their Google feedback form.

Scan this QR code to access a list of suggested materials for this activity!

You may have noticed that the speed at which a group works depends on the complexity, uniqueness, creativeness, and innovativeness of the solutions they have designed.

What are some ways you can motivate your students to explore more complex, unique, creative and/or innovative ideas?

Classroom Tidbit
Students may find it difficult to devise unique, creative, or innovative solutions. One way to overcome this is to hold regular brain dumping sessions to freely explore any unused idea of interest over a short period of time. You might like to try this at the end of an activity as part of the reflection process.

The feedback or evaluation process can be quite daunting for students. Aside from anonymity, in what ways can you make this process more enjoyable and/or meaningful for your students?

Classroom Tidbit
You might like to try a more interactive tool such as a live poll.

Part Three:
Creation & Testing of the Prototype

In part three of the integrated STEM activity, students are given another two hours to create, test, and evaluate their prototype. Once again, they have 15 minutes to plan how they will manage their time. However, this time, Mr Chen informs the students that they should also include time to evaluate, redesign and retest their prototypes, listing the various steps to help them organise their plans.

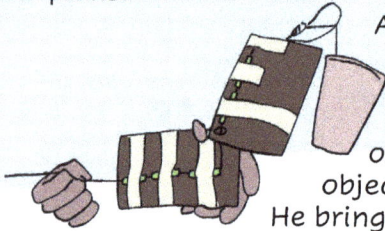

As the students construct their prototypes, Mr Chen notices that the group constructing a crane-like device to lift an object seems to only test their prototype with light objects, such as paper cups.

He brings a small bottle of water over to the group and reminds them that it is important to consider the types of objects that a patient might need to retrieve while resting in bed. He prompts them to consider how they can modify their prototype to lift heavier objects. Upon being met with stumped expressions, Mr Chen grabs a crate of materials and places it on their table. He guides his students to find a solution through stepwise questions.

For example:
- What happens to your prototype when you lift the small bottle of water, why?
- Which of the materials in this crate can make your prototype stronger so that it can lift a heavier object?

Group problem solving

Engaging in iterative design

This prompts the students to sample different materials to reinforce and improve their crane-like device. To their relief, they find ice cream sticks twice the width of those sold commercially. This refuels their motivation to continue the problem-solving process.

Making connections

Once the students have completed their prototypes, Mr Chen hands out a worksheet designed to help them think critically and reflect on their design and prototyping process.

He explains that,

"Designing a product can be a complex process that involves plenty of trial and error and change. It is important to reflect on what you have designed, who you are designing for, and what your users' needs are."

Scan this QR code to access the design thinking worksheet!

Take a moment to reflect on the integrated STEM activities you have conducted in class.

(a) What are some common challenges your students have faced during the creation, testing, evaluation, and redesign process?

(b) What are some exercises or lessons you can implement in the future to help them overcome similar challenges in the future?

Classroom Tidbit
Depending on the activity, you may wish to introduce students to various skills prior to starting a STEM activity. These skills can range from drilling to coding.

Part Four: Prototype Presentation & Evaluation

Communication

Teacher evaluation

Prioritising evidence

In the last part of the integrated STEM activity, students are given 10 minutes to present and five minutes for their classmates and teacher to evaluate the prototype. A representative from each group draws a number from a bag. The number represents the order in which the groups will give their presentations.

Students are evaluated based on three criteria:

- the function of the prototype (**meeting user needs**)

- the **science or evidence** used to support their design, and

- the **user friendliness** of the prototype.

As his students are relatively young, Mr Chen is not surprised when more than one group cannot produce a working prototype at the time of presentation. He is quick to reassure them that some ideas may require further work and cannot be completed within the given timespan. He ensures to cheer them on and praise them for their efforts. For groups that cannot complete their prototype, Mr Chen takes the time to ask about the remaining parts.

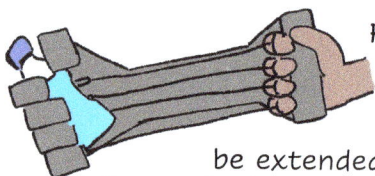

For example, when asked about their robotic arm, his students explain that they did not know how to make it "move like an actual robot" that can be extended and worn like a sleeve on John's arm. When asked to elaborate further, his students share that they were inspired by superheroes from movies and comics and wanted to make a superhero's "armour" for John's hand.

Recognising that other groups may have experienced similar limitations, Mr Chen holds an impromptu discussion at the end of the workshop for students to voice the ideas they wished to explore but were unable to due to a lack of resources or skills. Mr Chen can see that the students had creative and innovative ideas in mind. He reflected that with time and more exposure to other STEM-related skills such as block coding, the students can bring their ideas to life in future learning cycles.

Mr Chen was able to reflect on his students' creativity and innovation during their unplanned discussion at the end of the workshop.

(a) Have you had a similar conversation with your students? If given the opportunity, what questions would you ask them?

(b) How can creativity and innovation be assessed in the classroom? What methods have you used or would like to try?

A scoring or scale system can be used to assess different criteria and/or skills. They give us a way to measure aspects of learning such as content knowledge. Can creativity and innovation be measured in the same way?

Scan this QR code to access the student workbook for this activity.

Mr Chen's Instructional Model

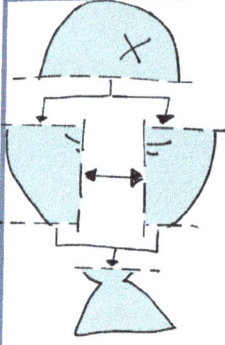

CRITICAL THINKING

<u>R</u>ecognise assumptions
<u>E</u>valuate arguments
<u>D</u>raw conclusions

Broad Learning Objective

Real-World Problem Statement or Scenario

Instructional Process

Support System
(Teacher Materials & Instructions)

Social System
(Student Activities)

Scan this QR code to access the detailed and editable instructional model.

[Adapted from] The Development and Validation of the Blended Socratic Method of Teaching (BSMT): An Instructional Model to Enhance Critical Thinking Skills of Undergraduate Business Students (3Q),1, p. 85 by E.A. Boa, A. Wattanatorn & K. Tsaong, 2018. Kasetsart Journal of Social Sciences. Copyright year 2023, by Kasetsart University.

Instructional Strategies

Students will:

Grade 11–12

Page 129 it is!

- Understand how some diseases are sex-linked and genetically inherited
- Understand how electrical muscle stimulation can be used to improve muscle strength and function
- Learn about resistance and endurance training
- Design a device to help a patient diagnosed with Duchenne muscular dystrophy improve lower limb function through one or more types of muscle training

> **Classroom Tidbit**
> Unsure of how to adapt Mr Chen's neuromusclar system activity for students of different grade levels? Pick your grade level, then follow the instructions.

Grade 9–10

Skip to page 128!

- Understand how genetic mutations can cause diseases
- Understand that some diseases are genetically inherited
- Understand how electrical muscle stimulation can be used to treat some neuromuscular diseases
- Design a device to help a patient diagnosed with limb-girdle muscular dystrophy to improve limb muscle strength and move around

Grade 7–8

Off to page 127!

- Learn about the neuromuscular system and how it works
- Understand how electrical muscle stimulation can promote oxygen and nutrient delivery to muscle fibres
- Understand how coronary artery disease can lead to stroke
- Design a single device to help a stroke rehabilitation patient increase muscle strength and perform specific fine movements

Grade 5–6

Flip the page!

- Learn about the neuromuscular system and how it works
- Understand how one system in the human body can affect another
- Learn about what stroke is and how it can affect movement
- Design two devices to help a stroke rehabilitation patient perform two specific fine movements

Grade 3–4

Back to page 110!

- Learn about the neuromuscular system and how it works
- Learn about what stroke is and how it can affect movement
- Design a device to help a stroke rehabilitation patient regain movement in his arm

At this level, students are aware that the brain is the control centre that coordinates and connects the systems in the body. The probe activity can be used as an analogy for how the neuromuscular system works. Students can then use their knowledge on the circulatory system to explain how a clot or rupture of a blood vessel in the brain can lead to neuromuscular impairment. Students can then be tasked to carry out a modified version of the featured STEM activity.

Task: John has been hospitalised with a mild stroke. He has lost muscle strength in the right side of his body and is experiencing muscle stiffness, particularly in his right arm. As a rehabilitation engineer, you have been tasked to assist John with his recovery. How can you develop a device that can help John perform tasks, such as feeding himself, turning off a light switch, or writing a note?

Students can develop and compare two different prototypes to perform two chosen tasks. The challenge is designed for students to consider the complex range of functions the neuromuscular system controls and appreciate the importance of user-centricity in product design.

What would you adopt or adapt from this activity to suit the learner profiles in your classroom?

At this level, although focus is still placed on the human digestive system and circulatory system, students learn about the concept of diffusion. They can appreciate and explain how electrical muscle stimulation in the probe activity can promote blood flow, oxygen, and nutrient delivery to the muscle fibres. As students also have knowledge of the interacting systems in the body, the featured STEM activity can be modified as follows.

Task: John has a history of coronary artery disease. He has recently been hospitalised with a mild stroke and is experiencing muscle weakness and stiffness in his dominant arm and hand. As a rehabilitation engineer, how can you develop a device that can help John increase his muscle strength over time and perform tasks, such as feeding himself, turning off a light switch, and writing a note?

In this task, students can be asked to explain how coronary artery disease can lead to the consequences of a stroke. They can be challenged to build one prototype that enables John to perform all three tasks.

What would you adopt or adapt from this activity to suit the learner profiles in your classroom?

For the grade 9-10 classroom

At grade 9—10, students are introduced to the concept of genetic mutations and inheritable traits and diseases. The probe activity can be used to introduce how electrical muscle stimulation (EMS) can be beneficial in the treatment of some neuromuscular diseases, such as the genetically inherited, limb-girdle muscular dystrophy (LGMD). The featured STEM activity can thus be modified to accommodate this context.

Task: Limb-girdle muscular dystrophy (LGMD) is a genetically inherited neuromuscular disease. It is characterised by progressive muscle weakness and atrophy in the hip and shoulder areas that may spread to other areas of the body. John is 43 and has been diagnosed with late onset LGMD. He experiences weakness in his thigh muscles and has difficulty getting up when seated. As such, he is often reliant on his arms to prop himself up. As a rehabilitation engineer, how can you develop an assistive device that can help John improve muscle strength in his limbs and move around?

In this task, students can be challenged to explore current rehabilitation methods and develop a device that combines simultaneous arm and limb training.

What would you adopt or adapt from this activity to suit the learner profiles in your classroom?

At this level, students learn about dominant and recessive genes, homozygosity, heterozygosity, and how genes are inherited. The probe activity can be used to introduce how electrical muscle stimulation can improve muscle strength and function through resistance training in some neuromuscular diseases. The featured STEM activity can hence be modified to the above concepts.

Task: Duchenne muscular dystrophy (DMD) is characterised by muscle weakness typically starting at the lower external muscles, progressing to the upper external muscles and loss of muscle mass. Structurally, zebrafish skeletal muscle is similar to human skeletal muscle. A recent study has shown that endurance training using neuromuscular electrical stimulation can improve zebrafish skeletal muscle structure, function, and survival. As a rehabilitation engineer, how can you design a device that could potentially help a DMD patient improve their lower limb function through endurance training?

In this task, students can learn about DMD and why it predominantly affects males. They can compare endurance training to other types of muscle training. Students can be further challenged to create a device that addresses more than one type of muscle training. Advanced leaners can explore concepts such as the reflex arc, neuronal and neuromuscular junction signalling and muscle contraction (crossbridge cycling).

What would you adopt or adapt from this activity to suit the learner profiles in your classroom?

Reflection Questions

As you approach the end of this book, take a moment to reflect on the integrated STEM activities you have enacted in your classroom.
(a) What part(s) of the prototype design process have you found most difficult to manage and why?

(b) What are some strategies you can implement to manage future STEM activity cycles more efficiently?

(c) How confident are you to carry out this activity in your classroom? Colour the fish to represent your confidence level and note down some reasons why.

Scan this QR code to access Mr Chen's editable lesson plan.

ACTIVITY SIX
THE CARDIOVASCULAR SYSTEM

Is your blood pressure considered healthy?

Probe Activity
What is Hypertension?

This probe activity centres on the context of cardiovascular disease in the ageing heart. It is designed to gauge students' prior knowledge on the cardiovascular system and its associated conditions.

Scenario

Aunty Mei is at the clinic to see a doctor. She has been experiencing headaches, shortness of breath, and at times, chest pain. Upon hearing her symptoms, the doctor decides to take her blood pressure.

133

Have you heard of the term "hypertension"?
Can you deduce what it means based on the
conversation above?

Classroom Tidbit
Lower grade level
students may not
be familiar with
the concept of high
blood pressure.
You may wish to
demonstrate what
happens to the
speed and pressure
of water when the
opening of a hose
is partially blocked
after the tap is fully
turned on.

What could happen if Aunty Mei's hypertension is left undiagnosed
over time?

Classroom Tidbit
Try using an analogy
to guide lower grade
level students to
understand how
chronic hypertension
can lead to a
condition such as
heart failure. Students
can be asked to
imagine the heart as
a motorised pump.
They can then be
asked to consider
what would happen if
the pump had to work
significantly harder
to pump blood around
the body over a long
time?

Why do you think Aunty Mei has to monitor her blood pressure daily?

How can regular exercise help improve Aunty Mei's blood pressure or general heart health?

The table below lists some relevant science concepts that can be covered in this integrated STEM activity on coronary artery disease and atherosclerosis.

Science Concepts/Grade Level	3—6	7—10	11—12
Anatomy and function of the heart	√	√	√
Arteriosclerosis and atherosclerosis			√
Atherectomy, angioplasty and stenting			√
Blood pressure regulation (baroreceptors)			√
Heart health (cardiovascular disease) and ageing	√	√	√
Pulmonary and systemic circulation	√	√	√
Structure and function of blood vessels		√	√
Vascular capacitance and blood pressure			√

What other science concepts or learning outcomes would you include in your lesson?

Implementing Integrated STEM Inquiry in the Grade 9—10 Classroom

Vignette

The vignette on the next page offers insight into how, "What is hypertension?" can be used to design a STEM activity on blood vessel structure and function.

The vignette describes how Mr Sun uses the 6E Instructional Model and the "Predict-Observe-Explain" strategy to implement integrated STEM inquiry in his classroom.

Integrated STEM Activity		
Part	Task	Time (minutes)
One	Engage, Explore, & Explain	120
Two	Elaborate, Evaluate, Extend, & Re-Evaluate	240

The featured integrated STEM activity requires students to (1) design and/or conduct experiments to determine the effects of vessel narrowing and stiffening on blood flow, (2) devise tools to conduct an atherectomy, and (3) investigate why the pattern or design of a stent is important. Included in the description are examples of his students' responses and his reflections on the process. The characteristics of STEM inquiry relevant to his lesson are noted in the margin on the left.

Scan this QR code to watch a video on cardiovascular disease.

Engage

Typically, when Mr Sun's students enter the classroom, their teacher is already waiting. Today, his students are instead greeted by a note saying, "Read 'What Is Hypertension?' in your workbook and think about the questions given. I will be back in 15 minutes."

To his students' bemusement, Mr Sun enters the classroom after 15 minutes with another teacher, Mr Balakrishnan, dressed in a lab coat with a stethoscope hung around his neck. After greeting the class, they begin a skit, with Mr Balakrishnan acting as Aunty Mei from the probe activity and Mr Sun as Dr Lim. Mr Sun has adapted the skit to weave in one of the probe activity's questions.

Mr Balakrishnan faces Mr Sun and asks, "Dr Lim, what would have happened if I didn't feel any symptoms and my hypertension goes undiagnosed for many years?" Here, Mr Sun has scripted their role play to use inquiry at the end of the skit to prompt his students to think more deeply.

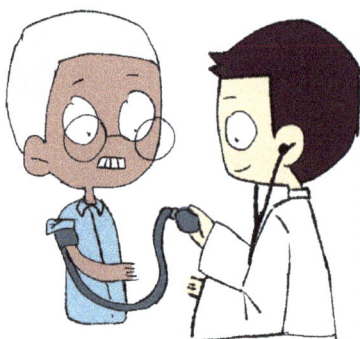

> Teacher-directed scientific inquiry

After the skit, Mr Sun thanks Mr Balakrishnan and gives his students 15 minutes to work individually, allowing his students to use their personal devices to find answers to the probe activity questions. Then, he groups his students in pairs to role play the probe scenario with the activity questions.

After five minutes of role play, students swap roles and repeat the skit again. Mr Sun has planned this activity to encourage his students to think more critically about the information they have found and to view things from different perspectives, learning from each other through the advice given as Dr Lim.

Next, Mr Sun presents the students with a challenge.

Integrated STEM Activity

Justin has coronary artery disease (CAD) and is scheduled for an atherectomy. As his daughter is only 10 years old, he wishes to design an investigation to help her better understand his situation. How can Justin design an investigation to explain (1) the effect of atherosclerosis on blood flow in CAD, and (2) what is used to conduct an atherectomy?

As Mr Sun has previously carried out problem-based integrated STEM activities at the end of each semester, his students are able to delineate the key problems presented from the context with ease. To challenge his students, he reveals that they must incorporate an automated component in the design of at least one of their investigations and they must use recyclable materials as much as possible.

Role play is a form of experiential learning. In this vignette, Mr Sun's students can take ownership of what they have learned by acting as Dr Lim. Have you tried to infuse role play into your teaching? What are your thoughts on implementing this teaching strategy in your classroom?

Classroom Tidbit
Role play is a useful tool to encourage students to interact, connect and communicate with each other. You may wish to bring a medical stethoscope to class to stimulate the students' sense of curiosity, and help them gain confidence when adopting the role of Dr Lim.

Explore

Mr Sun divides his students into groups of six. Rather than having his students deep dive into the activity, he presents them with a schematic of the circulatory system, which he calls the Circulatory Map. His students are given 10 to 15 minutes to label the map. After receiving requests for help from several students, he writes a list of helping words on the whiteboard to assist them. Mr Sun uses this exercise to tap his students' prior knowledge as they attempt to recall and reconstruct their knowledge of the circulatory system.

He follows the activity with a slide on blood vessel structure and function, reminding them of the three types of blood vessels: arteries, veins, and capillaries.

Making connections To reinforce his students' understanding of the hierarchical organisation of blood vessels in the body, he invites his students to participate in an analogy-based activity. Students are tasked to illustrate how highways, main roads and side streets can be used to describe blood vessel structure and how different types of vehicles can be used to represent components such as the types of immune cells found in our blood.

Scan this QR code to access the Circulatory System worksheet!

While observing his students, Mr Sun makes two observations.

1 While some students easily relate the circulatory system to the analogy of a road map, other students need guidance to associate the different types of vehicles to the components found in blood.

2 His students also present their analogies in different ways. Some students draw road maps to houses and buildings of different sizes, representing different organ types, while other students use flow charts to demonstrate the hierarchical structure of blood vessels.

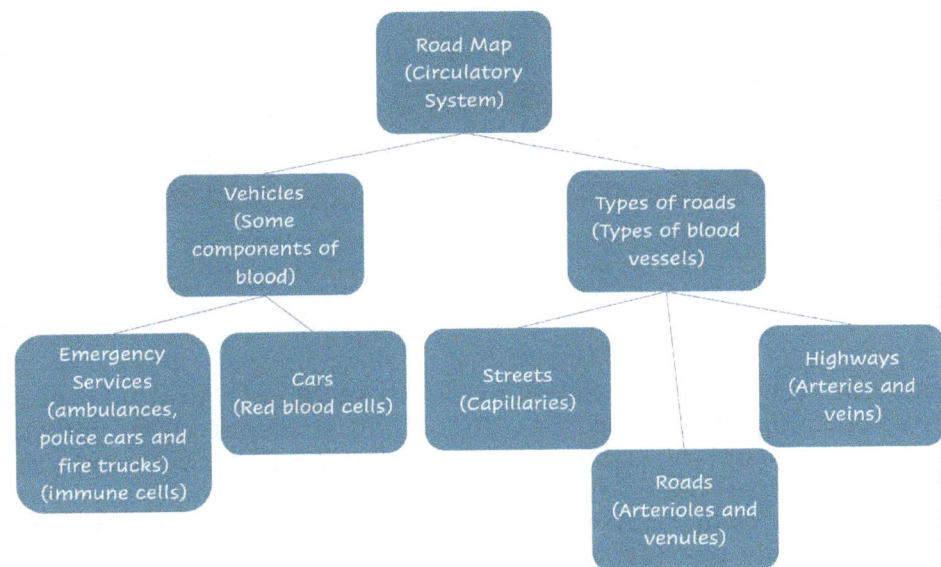

Road Map (Circulatory System)

Vehicles (Some components of blood)

Types of roads (Types of blood vessels)

Emergency Services (ambulances, police cars and fire trucks) (immune cells)

Cars (Red blood cells)

Streets (Capillaries)

Highways (Arteries and veins)

Roads (Arterioles and venules)

Explain

Group problem solving

Making connections

Guidance to identify the problem

Student-centric inquiry

Next, students are given time to work in their groups to **(1)** recall what they know about coronary artery disease (CAD) and **(2)** find out how plaque buildup in CAD can lead to adverse consequences such as a heart attack. Students then build on their road map analogy further. This time, Mr Sun invites his students to use the concept of traffic flow to describe the impact of atherosclerosis and arterial stiffening on the rate of blood flow in coronary artery disease.

While some students use a roadblock to describe how plaque accumulation can lead to arterial blockage, others use partial closure of a two-laned road to demonstrate how blood flow and blood supply is reduced with plaque buildup.

Roadblock (accumulation of plaque leading to a blocked coronary artery)

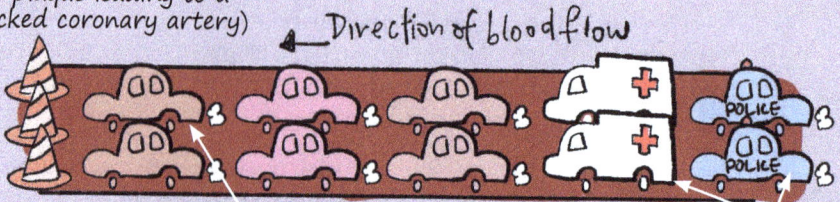

Direction of blood flow

Cars (Red blood cells) carrying oxygen cannot pass through the roadblock. The muscles that receive blood through this pathway will be deprived of oxygen and other nutrients. This can cause a heart attack and/or other consequences.

Other components of the blood such as immune cells cannot pass through the roadblock either.

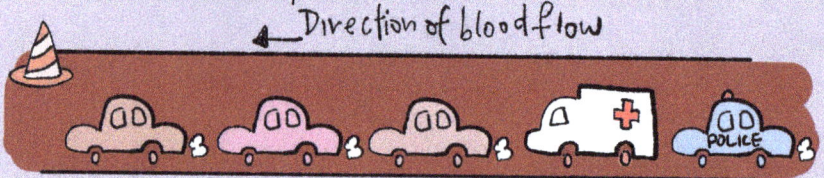

Direction of blood flow

Accumulation of plaque causes the artery to narrow

↓ Rate of blood flow

Plaque buildup can lead to narrowing of the coronary artery.

Mr Sun noticed that although some students initially found the road map analogy to be difficult, all students were able to link the concept of atherosclerosis and reduced blood supply to the concept of traffic flow. Upon reflection, he noted that his students enjoyed relating scientific concepts to contexts that they are likely to experience in their day-to-day lives.

Mr Sun uses two analogies to help his students understand the circulatory system and the effects of atherosclerosis on coronary blood supply. Which of these analogies would you adopt for your students and why? Alternatively, what other analogies would you use?

Part Two: Elaborate, Evaluate, Extend, & Re-Evaluate

Elaborate

Teacher-directed scientific inquiry

To begin this phase of the activity, Mr Sun writes the words "appendectomy", "hysterectomy", "vasectomy", and "lobectomy" on the whiteboard. Turning to his students he asks, "What do these words have in common?" The students are quick to respond, identifying that the words end in 'ectomy'.

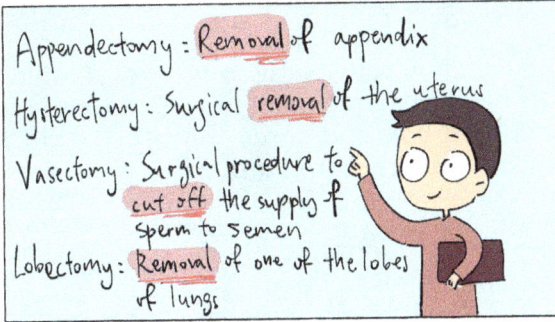

> Appendectomy: Removal of appendix
> Hysterectomy: Surgical removal of the uterus
> Vasectomy: Surgical procedure to cut off the supply of sperm to semen
> Lobectomy: Removal of one of the lobes of lungs

Mr Sun follows up with a second question, "Do you think 'ectomy' would have the same meaning in each of these scientific words?" This time, the responses varied, with some students remarking, "maybe" and "it depends". Upon hearing this, he asks them to find the meaning of each term, and selects volunteers to write them on the board.

Before the students can finish writing the definitions on the board, Mr Sun hears the correct answer from the onlooking students. He reinforces their observations by underlining the common key word(s) in each definition. He then asks, "Given what you see on the board, what do you think 'atherectomy' means?" Having done some research on atherosclerosis, his students are able to deduce that atherectomy is the removal of plaque from the artery (blood vessel).

Student-centric inquiry

Group problem solving

Happy with their responses, Mr Sun tasks the students to individually research what an atherectomy involves and the types of devices used to conduct it. After 45 minutes, Mr Sun gathers the students in their respective groups and divides each group into subgroups of three, where one subgroup designs an investigation into how atherosclerosis affects the blood supply to the heart in CAD, and the other subgroup designs and compares devices for an atherectomy.

Students are given one hour to research and brainstorm ideas. As different groups will approach each investigation differently, Mr Sun ensures to provide group-specific guidance to help the students **(1)** identify the purpose of the investigation, and **(2)** create an appropriate investigation plan. In particular, he focuses on guiding them towards solutions that are feasible given the time frame and resources. For example, some subgroups may require electrical devices that the school does not possess to automate the flow of water (i.e., blood flow) in their investigation on atherosclerosis. In anticipation of this, Mr Sun has prepared mini water pumps as a similar alternative.

To encourage his students to use as many recyclable materials as possible, Mr Sun holds a scavenger hunt. Student representatives are given a cardboard box and three minutes to fill the box with recyclables that he has scattered around the classroom.

> What about using straws of different diameters to represent an artery before and after plaque accumulation, if we cannot find any more tubing?

Group problem solving

Students can fill their boxes to the brim; however, they must ensure that each scavenged item is used in their investigation. To his delight, the scavenger hunt boosts his students' energy levels and fuels their sense of determination as they brainstorm how to incorporate each recycled item into the design of their investigation.

Group problem solving

Students then set about actualising their investigation plans. While constructing the various components required for their investigations, Mr Sun receives many requests for a second scavenger hunt. As the students already have an idea of the types of materials available, he asks them to make a list of what they need with suitable alternatives should the materials be taken by other groups.

He then holds a two-minute "first come first serve" scavenger hunt to challenge and motivate his students to think critically and quickly as both the materials they listed, and their suitable alternatives, are limited in volume and quickly being taken by other groups.

> Scan this QR code to access the list of materials for this activity!

144

Mr Sun's invites his students to deduce the meaning of atherectomy from a list of words with the same suffix. This may be difficult for lower grade level students. What are other ways you can introduce the concept of an atherectomy to lower grade level students?

Classroom Tidbit
You may wish to have students trade and barter materials with each other.

Mr Sun uses scavenger hunts to encourage his students to think creatively when incorporating the recyclable materials gathered. What other way(s) can students source the materials they need from their surroundings and be inspired to use them creatively?

Teacher-directed scientific inquiry

As the students continue constructing their investigation models or devices, Mr Sun reminds them to "Predict-Observe-Explain", a strategy he often uses in class for an investigation. This requires the students to hypothesise what they would expect to see based on the scientific knowledge used to inform the design of their investigation, making observations when conducting the investigation, and using scientific evidence to explain their conclusion. Students record their prediction (hypothesis), observations, and results in a guided laboratory notebook.

A total of 90 minutes is given to build and test their investigation model or atherectomy devices. As his students have participated in previous integrated STEM activity cycles, Mr Sun takes a step back to observe them. He reflects on their progress since the first semester and notes that they are now able to work more independently and have become more resourceful.

Scan this QR code to access a sample lab notebook.

Atherosclerosis Investigation Subgroups

During his observations, Mr Sun notices several of his students communicating openly, with a sense of ease, acceptance, and flexibility, taking turns to voice various ideas to overcome obstacles such as water leakage. Their suggestions are also supported by logical explanations. For example, when one student suggested to "place a layer of molding clay on the inner rim of the hole so that the tube will stay firmly in place when [they] push it through", another responded with, "yes, let's also put some clay around the outside of the tube to make sure all holes are sealed".

Scan this QR code to access how this investigation can be automated.

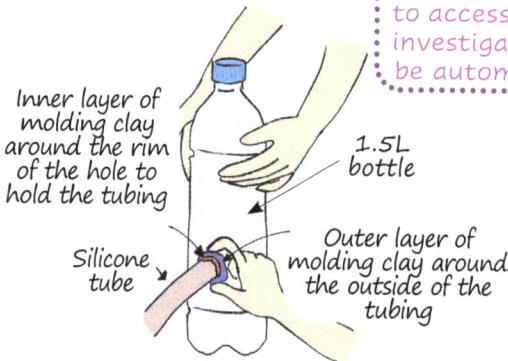

Inner layer of molding clay around the rim of the hole to hold the tubing

1.5L bottle

Silicone tube

Outer layer of molding clay around the outside of the tubing

Atherectomy Device Investigation Subgroups

Group problem solving

Prioritising evidence

Mr Sun makes similar observations with the atherectomy device subgroups. As students must use all the recyclable materials obtained, he can see their efforts in ensuring that each material has a purpose. For example, when modeling their device after the Jetstream Atherectomy System's motorhead, one group decides to revise their initial plan of using their own coins to form the rotational blades. Instead, they use metal bottle caps they received when trading materials with another group. They rationalise that "the metal bottle caps are better because they have grooves that stick outwards to remove the plaque more effectively."

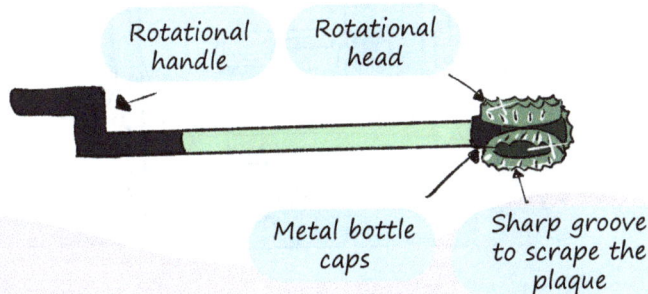

Rotational handle

Rotational head

Metal bottle caps

Sharp groove to scrape the plaque

Making connections

In addition, compared to the previous integrated STEM activity cycles, Mr Sun notices that his students can connect various facets of their solution with the content knowledge and/or skills from the disciplines of STEM more easily than before. For example, while making his rounds, he chances upon a student measuring the angles of the blades on a screwdriver head. When asked to elaborate, the student replies, "This screwdriver is designed for one type of screw. The blades are angled this way on purpose. I am measuring the angle so that we can position the metal caps in the same way. We want our device to look like and function almost like a screwdriver."

Similarly, another student was quick to deduce that using appliances like a mini pump to regulate water flow is one way of incorporating technology into their investigation. Mr Sun reasons that the regular end-of-semester integrated STEM activities have played an integral role in providing guidance and opportunities to foster an integrative mindset.

Evaluate

Next, Mr Sun invites the subgroups to present their prototypes and evaluate each other. During the presentations, Mr Sun notices that the atherosclerosis subgroups experienced difficulties in automating their investigation models. Upon closer inspection, Mr Sun realises that in their haste to complete their models, the students have missed some key details. For example, compared to the model on the right, the 'blood' flows more quickly out of the outlet straw (blood vessel) filling the container below in the model on the left. This is because the suction straw used is much wider than the outlet straw. The suction straw must be smaller in diameter than the outlet straw.

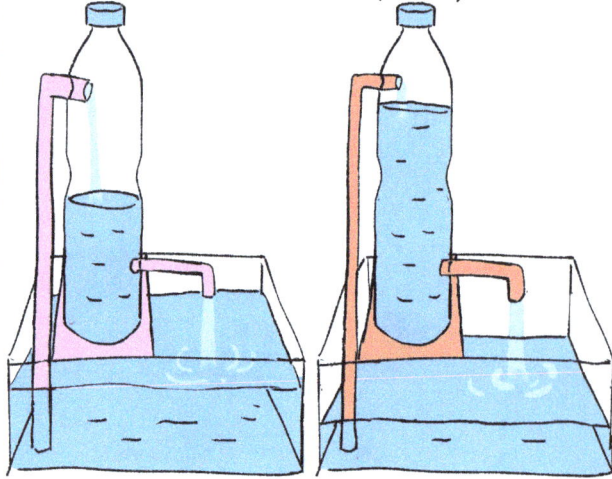

Mr Sun was about to reflect on his students' progress, having previously conducted a few integrated STEM activities with his class. What signs of progress or growth have you noticed in your students' STEM learning journey?

You may have noticed that the students are not always successful in building a working prototype. As the purpose of these integrated STEM activities is to develop students' 21st century skills, STEM knowledge and competencies:

(a) would a theoretical idea on how a prototype could be automated suffice as a demonstration of creativity and innovation, and why?

(b) how much value should a judge or evaluator put towards the delivery of a working prototype and why?

For the atherectomy device investigation subgroups to be fairly assessed, Mr Sun prepares equal sized tubes with equal amounts of clay placed in similar positions along the inside of the tube, to represent arteries with plaque buildup.

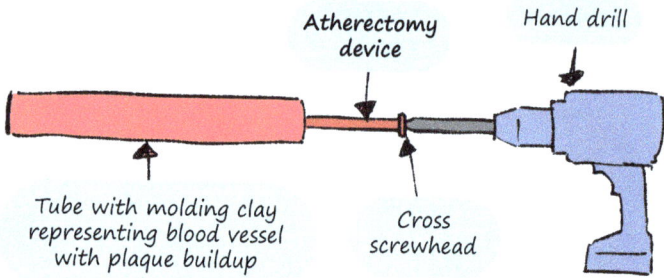

Atherectomy device

Hand drill

Tube with molding clay representing blood vessel with plaque buildup

Cross screwhead

Mr Sun can see the creativity and innovativeness behind the devices created. For example, one group presented an atherectomy device powered by a drill. The group created three different atherectomy devices, each with a cross screwhead at its end, allowing the user to switch between them when needed, while using the hand drill to automate it and control the rate of movement (i.e., plaque removal). This meant that they could conduct a fair test to determine which of the three devices could remove the most plaque.

Students also used investigation-specific Google Forms set up by their teacher to evaluate each other.

For example, the atherectomy device Google Form includes questions such as:

- Is the prediction or hypothesis supported by the group's observations?
- Has the group used science-based evidence to design the device?
- How effectively does the best device remove the plaque in the artery?
- Can the device consistently remove the same amount of plaque?

Google Forms also enables Mr Sun to immediately view the distribution of the students' responses.

Study the atherectomy device designed by the group. How effective is the device in removing plaque?

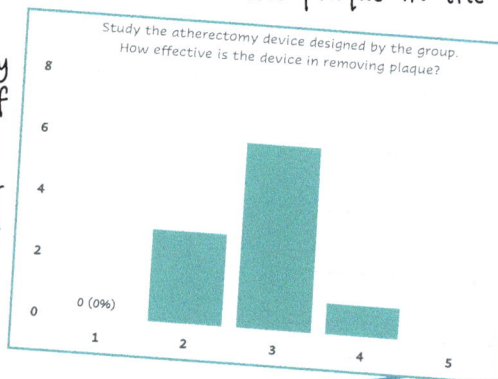

In this vignette, Mr Sun took care in preparing the arteries (tubes) with plaque (molding clay) deposits, such that the atherectomy devices can be fairly assessed.

What are some variables you can control to ensure that the:

(a) atherosclerosis investigation subgroups are fairly evaluated?

> **Classroom Tidbit**
> One way you can ensure that all atherosclerosis investigation subgroups are fairly evaluated is to have students start with the same volume of water. You may wish to adjust the volume of water for models that use larger containers.

(b) atherectomy device investigation subgroups are fairly evaluated?

Extend

Conscious of the time, Mr Sun does not conduct a second cycle of explore, elaborate, and evaluate as he had initially planned. Instead, he modifies his plan to include an extend and re-evaluation phase, so that students can still engage in iterative design. Here, students are given a chance to improve both their investigation model and device prototype(s). Mr Sun intentionally reunites the six-member groups to include the involvement and views of more students. He hopes that in doing so, he can simulate the prototype/product evaluation process that occurs in the industry, as external stakeholders, designers, or researchers are oftentimes brought in to evaluate and at times, help improve a prototype/product.

Engaging in iterative design

Group problem solving

Students are given 60 minutes to improve their investigation model and atherectomy devices based on the feedback from Mr Sun and their peers. Mr Sun also reminds his students to "ensure that [they] are making fair comparisons. For example, if [they] are comparing arteries (straws) of two different diameters, then [they] must control for variables such as the length of [their] straws."

Given the limited time, not all groups can complete the improvements they intended to make. Mr Sun reassures them that the prototyping process can often take more than several hours and that it is just as important to be able to explain the improvements they intend to make and why.

Student-centric inquiry

For students that have completed their investigation models and atherectomy devices, Mr Sun challenges them to investigate how the design or geometric pattern of a stent can affect its function. Before beginning their research, students must brainstorm the factors that a biotechnology company should consider when designing a stent and how the factors may affect the delivery, use, and function of the stent.

Depending on the amount of time left, students could also sketch and present their own stent design during the re-evaluation phase.

In the extend phase, Mr Sun intentionally reunites the six-member groups to mimic the prototype/product testing and evaluation process that could occur in the industry.

(a) Would you consider adopting this approach for your classroom and why?

(b) How would you do this differently from Mr Sun?

Re-Evaluate

Teacher evaluation

Next, students present their (1) revised atherosclerosis investigation model, (2) atherectomy device(s), and (3) stent research findings and design sketches (optional) to a panel of five judges. Of the five judges, four of them are teachers representing each of the STEM disciplines, while the last judge is a music teacher representing a neutral party.

Overall, students are able to improve at least one aspect of each prototype. For example, although one group manages to automate their atherosclerosis model using a mini water pump, they cannot solve the newly presented problem of achieving equilibrium, such that one container would continue to overfill while the other container would be drained of 'blood'.

Water about to overflow

Mini water pump

Feedback from the judges vary. For example, in his evaluation, the design and technology teacher suggests that the students should consider using Arduino to control the mini water pump. In this way, they incorporate the use of technology in their investigation and develop their digital literacy skills at the same time. On the other hand, the mathematics teacher proposes to incorporate more mathematical skills, such as creating a proportionally larger device to scale to one that is used to conduct an atherectomy.

Communication

As most of the groups could not reach the "design your own stent" stage, Mr Sun decides to not have this component evaluated.

However, he encourages the students to present their findings in the form of a sharing session. In general, the students share similar findings, that a reduced number of connectors increases the flexibility of a stent but can also cause longer stents to become deformed when compressed.

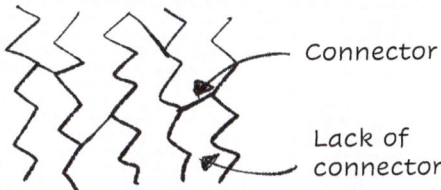

Connector

Lack of connector

Lastly, Mr Sun gives his students time to reflect on the integrated STEM activity, to provide feedback on what they like and what they would change or add to improve their STEM learning experience. As expected, many students want to have more time, expressing that the activity could be conducted across two days to allow them to "make [their] prototype better" or "find a better way to use technology in [their] design".

Mr Sun gathers feedback from his students on their STEM learning experience at the end of the activity. Try holding a feedback session of your own, then reflect on how you can improve your students' learning experiences in the space provided.

Scan this QR code to access the student workbook for this activity.

Mr Sun's Instructional Model

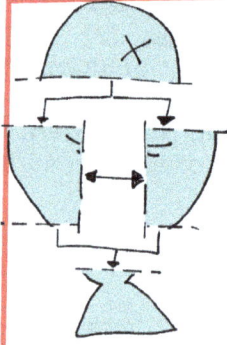

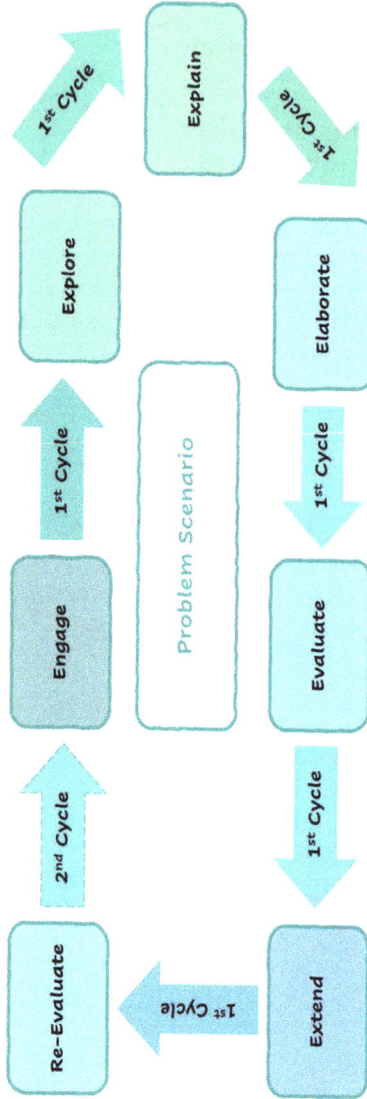

Engage

Explore

Explain

Elaborate

Evaluate

Extend

Re-Evaluate

Problem Scenario

1st Cycle

1st Cycle

1st Cycle

1st Cycle

1st Cycle

1st Cycle

2nd Cycle

Scan this QR code to access the detailed and editable instructional model.

Instructional Strategies

Classroom Tidbit
Unsure of how to adapt Mr Sun's activity on the cardiovascular system for students of different grade levels? Pick your grade level, then follow the instructions.

Students will:

Grade 11–12

Page 161 it is!

- Recall knowledge on blood flow through the heart
- Learn about atherosclerosis in coronary artery disease (CAD) and how it can be treated with an atherectomy and angioplasty with stent placement
- Design an experiment to investigate blood flow in CAD (include an automated component)
- Design an automated device to investigate what is used in an atherectomy
- Investigate how the design of a stent is important for its function
- Propose, create and compare different stent designs

Grade 9–10

Back to page 137!

- Recall knowledge on blood flow through the heart
- Learn about atherosclerosis in coronary artery disease (CAD) and how it can be treated with an atherectomy
- Design an experiment to investigate blood flow in CAD (include an automated component)
- Design an automated device to investigate what is used in an atherectomy
- Learn about angioplasty with stent placement
- Investigate how the design of a stent is important for its function (optional)

Grade 7–8

Off to page 160!

- Learn about the coronary blood supply
- Learn about atherosclerosis in coronary artery disease (CAD)
- Design an experiment to investigate blood flow in CAD
- Design a device to investigate what is used in an atherectomy

Grade 5–6

Skip to page 159!

- Learn about blood vessel structure and function
- Learn about atherosclerosis
- Design an experiment to investigate how atherosclerosis can affect blood flow
- Design an experiment to investigate why an atherectomy would be helpful

Grade 3–4

Flip the page!

- Learn how the heart functions
- Learn about the circulatory system
- Build a model of the heart

At this stage, grade 3—4 students have primarily focused on the digestive system. As they are not familiar with the cardiovascular system, you may wish to introduce them to a model of the heart and have them explore the various parts of the heart and their function. Students can then learn about how the heart connects the circulatory and respiratory systems through a simple activity, such as "Take a Walk Through the Circulatory Maze", in which students trace their way around the circulatory system to meet various characters (i.e., organs) supplied by blood from the four-room pump headquarters (i.e., the heart). Students can then be guided to carry out the following task.

Task: You are preparing for your school's annual Science fair. As you have recently learned about the human circulatory system, your teacher has tasked you to find out, "How can you build a model that shows the pathway of blood through the heart and to the rest of the body?"

In this task, students can be further challenged to learn about how the structure of the valves in the heart relate to their function. Although the students are not required to know the specific names of the valves at this stage, they can appreciate how impaired function in one of the valves (i.e., in valve diseases), can affect the direction of blood flow through the heart.

What would you adopt or adapt from this activity to suit the learner profiles in your classroom?

For the grade 5-6 classroom

In grades 5–6, students can identify the different parts and functions of the human respiratory and circulatory systems. They can appreciate how these systems are also integrated so that the organs of the body can receive the blood supply and oxygen they need. As such, students can learn about the different types of blood vessels in the body, and their hierarchical organisation. Students can then be challenged to find out more about atherosclerosis and how it can affect the delivery of blood, oxygen and nutrients to an organ. Students can then be asked to conduct a simplified version of the atherosclerosis investigation in the featured STEM activity.

Task: Justin has been diagnosed with atherosclerosis and is scheduled for an atherectomy. As his daughter is only 10 years old, he wishes to design an investigation to help her better understand his situation. How can Justin design an investigation to explain (1) the effect of atherosclerosis on blood flow and (2) why an atherectomy would be helpful?

Students can be challenged to learn about the coronary blood supply and how atherosclerosis in a coronary artery can lead to a heart attack or compromise heart function. Students can then be asked to propose how they would use their current investigation model to demonstrate a coronary occlusion.

What would you adopt or adapt from this activity to suit the learner profiles in your classroom?

At this level, students are familiar with the circulatory system and the respiratory systems. They can be asked to explain how the systems are linked. Students also learn about how the circulatory system is important in the diffusion of key substances, such as nutrients and oxygen to the organs in the body. Given this, students can find out about the coronary blood supply, coronary artery disease and how it is treated. Students can then be tasked to conduct the integrated STEM activity featured in the vignette, without the extended activity on stent design.

Task: Justin has coronary artery disease (CAD) and is scheduled for an atherectomy. As his daughter is only 10 years old, he wishes to design an investigation to help her better understand his situation. How can Justin design an investigation to explain (1) the effect of atherosclerosis on blood flow in CAD, and (2) what is used to conduct an atherectomy?

Here, students need not incorporate automation into their prototype designs. However, students can still be challenged to think creatively and innovatively by using as many recyclable materials as possible. For example, students can be given bonus marks as an incentive for incorporating various types of recyclable materials. The bonus marks can be given as part of a tiered system, such that students are awarded a different amount of bonus points based on the type and quantity of recyclable materials used.

What would you adopt or adapt from this activity to suit the learner profiles in your classroom?

By grades 11–12, students have learned about the circulatory system, blood vessel structure, and how atherosclerosis can lead to cardiovascular diseases, such as coronary artery disease. Although students may have a general idea of how atherosclerosis is treated, they do not possess detailed knowledge on how it is performed, and the devices that are used. Students can hence be challenged to undertake a modified version of the STEM activity featured in the vignette.

Task: Justin has coronary artery disease (CAD) and is scheduled for an atherectomy, followed by angioplasty with stent placement. As his daughter is only 10 years old, he wishes to design an investigation to help her better understand his situation. How can Justin design an investigation to explain (1) the effect of atherosclerosis on blood flow in CAD, (2) what is used to conduct an atherectomy, and (3) how the design of a stent can affect its use and function?

You may wish to challenge your students further by having them automate both their investigation model and atherectomy device. In addition, students can also be tasked to create and investigate how various stent designs can affect their uses and functions.

What would you adopt or adapt from this activity to suit the learner profiles in your classroom?

Reflection Questions

Now that you have reached the last integrated STEM activity in this book:

(a) How would you describe your STEM learning journey?

(b) What were some of your toughest challenges:
(i) as a teacher?

(ii) as a learner of integrated STEM teaching?

How confident are you to carry out this activity in your classroom?
Colour the fish to represent your confidence level and note down some
reasons why.

What are some other integrated STEM activities or topics you would like
to explore?

How confident are you to plan and carry out a new integrated STEM
activity of your own? Colour the fish to represent your confidence level
and note down some reasons why.

Scan this QR code
to access Mr Sun's
editable lesson plan.

QR Code Resources

Introduction:

The STEM Quartet Instructional Framework

Activity One: Urban Farming

1. Video on vertical farming
2. Suggested list of materials
3. Student workbook
4. Ms Yeo's instructional model
5. Ms Yeo's lesson plan

Activity Two: Rusting

1. Video on rusting
2. How to make rust fast
3. Worksheet (Physical and chemical changes)
4. Student workbook
5. Ms Kee's instructional model
6. Ms Kee's lesson plan

Activity Three: Oil Spill

1. Video on an oil spill at sea
2. Topic-related riddles
3. Suggested list of materials
4. Student workbook
5. Mr Williams' instructional model
6. Mr Williams' lesson plan

Activity Four: Test Kits

1. Video on test kits
2. Suggested list of materials
3. Checklists
4. Evaluation form
5. Student workbook
6. Mrs de Silva's instructional model
7. Mrs de Silva's lesson plan

Activity Five: The Neuromuscular System

1. Video on the neuromuscular system
2. PowerPoint slides (Neuromuscular system and disease)
3. Suggested list of materials
4. Worksheet (design thinking)
5. Student workbook
6. Mr Chen's instructional model
7. Mr Chen's lesson plan

Activity Six: The Cardiovascular System

1. Video on cardiovascular disease
2. Worksheet (Circulatory system)
3. Suggested list of materials
4. Lab notebook
5. Video on atherosclerosis prototype
6. Student workbook
7. Mr Sun's instructional model
8. Mr Sun's lesson plan

www.ingramcontent.com/pod-product-compliance
Lightning Source LLC
Chambersburg PA
CBHW060017220326
41518CB00042BB/2226